Designing Web APIs with Strapi

Get started with the Strapi headless CMS by building a complete learning management system API

Khalid Elshafie

Mozafar Haider

BIRMINGHAM—MUMBAI

Designing Web APIs with Strapi

Group Product Manager: Pavan Ramchandani

Publishing Product Manager: Aaron Tanna

Senior Editor: Keagan Carneiro

Content Development Editor: Adrija Mitra

Technical Editor: Shubham Sharma

Copy Editor: Safis Editing

Project Coordinator: Rashika Ba

Proofreader: Safis Editing

Indexer: Manju Arasan

Production Designer: Joshua Misquitta

Marketing Coordinator: Anamika Singh

First published: February 2022

Production reference: 1210122

Published by Packt Publishing Ltd.

Livery Place

35 Livery Street

Birmingham

B3 2PB, UK.

978-1-80056-063-5

www.packt.com

To Ihsan, Elshafie, Bushra, Azza, Zhang He, and Sami for always loving and supporting me. To Hadi, without whom this book would have been completed way earlier. To myself, well done.

– Khalid Elshafie

To Meriem, for her patience and support before, during, and after writing the book. To Hayat and Hayder, my beautiful parents, for getting us here against all odds.

To Sudan and its youth, bravely forging a brighter future for all of us.

– Mozafar Haider

Contributors

About the authors

Khalid Elshafie is an experienced, senior full stack developer/software engineer with over 10 years of experience working across multiple frontend and backend technologies involved in designing and developing scalable web and mobile applications using multiple technologies, as well as in a variety of workplaces, from start-ups to larger consultancies. Khalid's passion for software engineering extends to the creation of barmaga.io, with Mozafar, where he has 45+ hours of video content on JavaScript, React.js, AWS and Serverless, Node.js, and Strapi. He also has a well-established YouTube channel with content focused on teaching programming.

I wish to thank my loving wife and son for their continued support and encouragement throughout the long process of writing this book. Also, I would like to thank Mozafar for introducing me to the world of programming 20 years ago, and last but not least, special thanks to Rares Matei for his tremendous efforts in providing valuable technical feedback for this book.

Mozafar Haider is a senior full stack engineer with over 15 years of experience working in organizations ranging from early-stage start-ups to scale-ups and corporates, in Barcelona, London, and Glasgow, among other places. He's passionate about teaching coding, especially to groups under-represented in tech, and was one of the co-founders of a coding school for refugees based in Glasgow. He also created, along with Khalid, barmaga.io, a platform for teaching coding in Amharic, Arabic, and Swahili.

I would like to thank Khalid for getting me involved in this project in the first place. We've come a long way since trying to learn ASP.NET 20 years ago on a hot summer's day in Khartoum. And a special thank you for pushing through the final phases of the book when I became overwhelmed with other life events. Here's to another 20 years of exciting projects!

About the reviewer

Rares Matei works for Nrwl.io on NxCloud, a zero-config distributed computation caching solution, helping teams and clients to speed up and scale their development practices with an open source tool called Nx, while also advising clients on web development best practices. He enjoys learning by teaching and has devised Egghead.io courses on TypeScript, reactive programming, and GraphQL. He is an organizer of the GlasgowJS meetup and an occasional speaker at other meetups and conferences.

Table of Contents

3

Strapi Content-Types

4

An Overview of the Strapi Admin Panel

Section 2: Diving Deeper into Strapi

5

Customizing Our API

6

Dealing with Content

7
Authentication and Authorization in Strapi

8
Using and Building Plugins

Section 3: Running Strapi in Production

9

Production-Ready Applications

10

Deploying Strapi

11

Testing the Strapi API

Appendix: Connecting a React App to Strapi

Index

Other Books You May Enjoy

Preface

Strapi is a Node.js-based headless **Content Management System** (**CMS**). It is flexible, developer-friendly, open-source, and free. It saves API development time through an integrated admin panel that is user-friendly and customizable, but it is also built with flexibility and extensibility in mind to allow developers to easily extend their APIs. The APIs built with Strapi can be consumed from any client using REST or GraphQL.

This book provides a complete step-by-step explanation of the essential concepts of Strapi through practical examples, in the context of real-life scenarios and common requirements for API development. The book can be read from cover to cover or section by section, for those looking for help with a specific topic in Strapi.

You will learn what Strapi is and how it works. We will start by building a simple API from scratch. Then, we will add complex features to the API, such as user authentication, data sorting, and pagination. Finally, we will learn how to deploy the API to Heroku and AWS.

Who this book is for

This book is for backend and frontend JavaScript developers. Experienced API developers will learn a new, fast, and flexible way of building APIs, while frontend developers will be able to take a step toward becoming full-stack developers by learning how to leverage Strapi for building APIs quickly. Basic knowledge of JavaScript and REST API concepts is assumed.

What this book covers

Chapter 1, *An Introduction to Strapi*, starts with giving a general understanding of what a headless CMS is, as well as what Strapi is all about and its benefits.

Chapter 2, *Building Our First API*, explores the structure of a Strapi API project.

Chapter 3, *Strapi Content-Types*, teaches us about Strapi content-types and the Content-Type Builder plugin and how to use it to create the API contents.

Chapter 4, *An Overview of the Strapi Admin Panel*, gives an overview of the Strapi admin panel and the different operations you can perform from the admin panel.

Chapter 5, *Customizing Our API*, shows how to customize the API and understand the models, controllers, and services in Strapi.

Chapter 6, *Dealing with Content*, teaches us how to work with the API contents, filter and sort data, implement pagination, and work with complex queries

Chapter 7, *Authentication and Authorization in Strapi*, teaches us how authentication and authorization work in Strapi. We will learn how to implement login and signup and how to secure the API endpoint.

Chapter 8, *Using and Building Plugins*, provides an overview of the Strapi plugin system, how to install and use plugins, as well as creating a simple Strapi plugin.

Chapter 9, *Production-Ready Applications*, teaches us how to prepare the API for production and how to keep permissions in sync between multiple environments.

Chapter 10, *Deploying Strapi*, provides hands-on examples of deploying Strapi to Heroku and AWS Fargate.

Chapter 11, *Testing the Strapi API*, teaches us how to test the Strapi application using Jest.

Appendix, *Connecting a React App to Strapi*, shows an example of connecting a frontend application to the API we have built throughout this book. We have built a simple and minimal appl that allows students to log in to their account, browse different classrooms and view tutorials in each classroom, students also have the access to enroll in a classroom. The application is built with React.js using Typescript.

To get the most out of this book

You will need Node.js 16 (or later) installed on your computer. All code examples have been tested using Node.js 16 on macOS. However, they should work with future versions too.

Optionally, you can install Docker if you want to deploy the application as a Docker container to AWS Fargate.

Software requirements: Node latest LTS, yarn (or npm), and Docker (optional)

Operating system requirements: Linux, MacOS or Windows

Download the example code files

You can download the example code files for this book from GitHub at `https://github.com/PacktPublishing/Building-APIs-with-Strapi`. If there's an update to the code, it will be updated in the GitHub repository.

Using this repository

The `main` branch has the latest state of the project. Use `tags` to access a specific chapter's code:

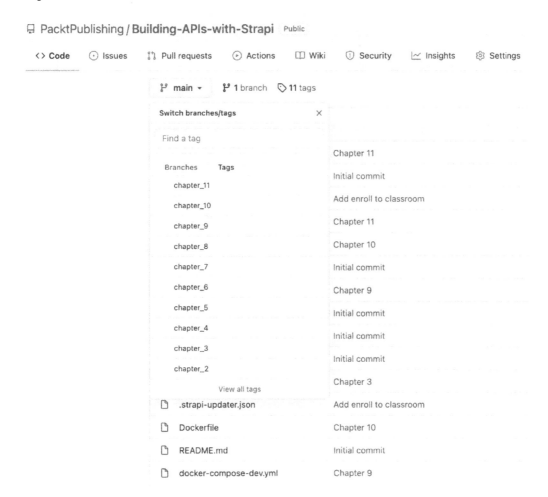

Following along with the book

If you would like to use this repository as a starting point, you can create a branch from `chapter_1` and follow along with the book from there:

1. **Clone the repository:**

   ```
   git clone git@github.com:PacktPublishing/Building-APIs-
   with-Strapi.git
   ```

   OR

   ```
   git clone https://github.com/PacktPublishing/Building-
   APIs-with-Strapi.git
   ```

2. **Navigate to the project root folder:**

   ```
   cd Building-APIs-with-Strapi
   ```

3. **Create new branch from chapter_1:**

   ```
   git checkout BRANCH_NAME chapter_1
   ```

4. **Install the dependencies:**

   ```
   yarn install
   ```

5. **Start the development server:**

   ```
   yarn develop
   ```

We also have other code bundles from our rich catalog of books and videos available at `https://github.com/PacktPublishing/`. Check them out!

Download the color images

We also provide a PDF file that has color images of the screenshots and diagrams used in this book. You can download it here: `https://static.packt-cdn.com/downloads/9781800560635_ColorImages.pdf`.

Conventions used

There are a number of text conventions used throughout this book.

`Code in text`: Indicates code words in text, database table names, folder names, filenames, file extensions, pathnames, dummy URLs, user input, and Twitter handles. Here is an example: "The `except` parameter is basically the opposite of the `only` parameter."

A block of code is set as follows:

```
async findTutorials(ctx) {
    ctx.response.status = 501;
    return "to be implemented";
}
```

When we wish to draw your attention to a particular part of a code block, the relevant lines or items are set in bold:

```
async find(ctx) {
  // Calling the default core action
  const { data, meta } = await super.find(ctx);
  meta.totalTutorials = data.length + 100;
  return { data, meta };
},
```

Any command-line input or output is written as follows:

```
docker push REPO_URL_HERE:latest
```

Bold: Indicates a new term, an important word, or words that you see onscreen. For instance, words in menus or dialog boxes appear in **bold**. Here is an example: "Since we are going to run a Fargate cluster, choose **Networking Only** from the cluster template and then click the **Next step** button."

> **Tips or Important Notes**
> Appear like this.

Get in touch

Feedback from our readers is always welcome.

General feedback: If you have questions about any aspect of this book, email us at customercare@packtpub.com and mention the book title in the subject of your message.

Errata: Although we have taken every care to ensure the accuracy of our content, mistakes do happen. If you have found a mistake in this book, we would be grateful if you would report this to us. Please visit www.packtpub.com/support/errata and fill in the form.

Piracy: If you come across any illegal copies of our works in any form on the internet, we would be grateful if you would provide us with the location address or website name. Please contact us at copyright@packt.com with a link to the material.

If you are interested in becoming an author: If there is a topic that you have expertise in and you are interested in either writing or contributing to a book, please visit authors.packtpub.com.

Share your thoughts

Once you've read *Designing Web APIs with Strapi*, we'd love to hear your thoughts! Scan the QR code below to go straight to the Amazon review page for this book and share your feedback.

https://packt.link/r/180056063X

Your review is important to us and the tech community and will help us make sure we're delivering excellent quality content.

Section 1: Understanding Strapi

Each journey starts with a first step. Our first step will be to understand the concept of headless **CMS** (short for **Content Management System**) and why Strapi is one of the most popular headless CMS around. Then, the chapters will move onto how to prepare the development environment and install and set up Strapi to build our first API.

In this section, we will cover the following topics:

- *Chapter 1, An Introduction to Strapi*
- *Chapter 2, Building Our First API*
- *Chapter 3, Strapi Content-Types*
- *Chapter 4, An Overview of the Strapi Admin Panel*

1

An Introduction to Strapi

This chapter gives an introduction to Strapi. First, we will explain what Strapi is and what are the benefits of using it in developing **API** (short for **application programming interface**). Then, we will see how to set up and prepare our development environment to work with Strapi. Afterward, we will create a simple API to help us get started with Strapi. Finally, we will have a quick look at the server scripts offered by Strapi to start and stop the development server.

These are the topics we will cover in this chapter:

- What is Strapi?
- Why use Strapi? (The benefits of Strapi)
- Preparing the development environment
- Creating a Strapi application
- Understanding server scripts

What is Strapi?

Strapi is a headless **content management system (CMS)**.

A CMS is a software application used for web development that allows users to create, edit, and publish content.

A traditional CMS, such as WordPress, tightly couples the frontend and the backend—that is, the structure of the content and how it is presented. However, unlike traditional CMSes, a headless CMS is entirely decoupled from the presentation layer. The term *headless* comes from separating the head (the frontend) from the body (the backend). A headless CMS does not care about how the contents get displayed; instead, it provides a content-first approach with an API to access and display the data in any format desired.

You can see the differences between a traditional CMS and a headless CMS in the following diagram:

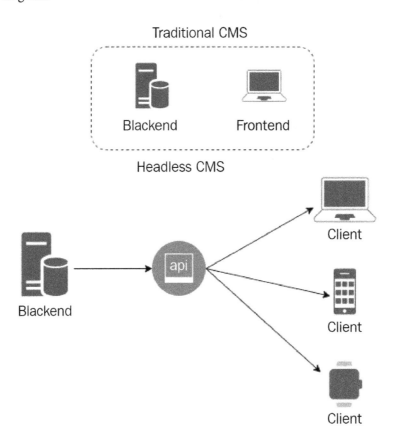

Figure 1.1: Traditional CMS versus headless CMS

"Strapi" is wordplay for **Bootstrap API**. As the name suggests, it aims to help developers build (bootstrap) an API quickly. Strapi saves API development time through an integrated easy-to-use admin panel and a solid set of core features out of the box. Whether you are a backend, full stack, or frontend developer, you will find it extremely easy to get started with Strapi and API development without reinventing the wheel and wasting time in building common features such as basic **create, read, update, and delete** (**CRUD**) operations or authentication and authorization.

Now that we have familiarized ourselves with Strapi, let's see some of the advantages of using it.

Why use Strapi? (The benefits of Strapi)

There are several advantages to using Strapi, and we will outline some of the major benefits and reasons for using it next.

Open-source Node.js

Strapi is open-source and built on top of a popular Node.js framework: Koa. Its code is accessible and easily extendible. It's supported by a thriving company but also by a large community of contributors. You can find Strapi's code on GitHub at `https://github.com/strapi/strapi`.

Database-agnostic

Strapi can work with different database systems. It can be set up and configured to work with PostgreSQL, MySQL, MongoDB, and SQLite. We will see in *Chapter 9*, *Production-Ready Applications*, how we can configure Strapi to work with PostgreSQL.

Customizable (extendable)

Strapi is highly configurable to suit each project's specific requirements. All of the data types are created from scratch through the admin panel. Additionally, the Strapi plugin system makes it easy to extend its functionality with features such as database documenting, image uploads, and email configuration. We will discuss the Strapi plugin system in a later chapter of this book.

RESTful and GraphQL

Strapi provides a **REpresentational State Transfer** (**REST**) API out of the box. The API can be consumed from any web client (React, Angular, Vue.js, and so on), mobile applications, or even **internet of things** (**IoT**) applications using REST or GraphQL via a plugin.

Users and permissions

Strapi comes with a users and permissions model out of the box that allows you to define which endpoint is available for which user/role. Additionally, you can use **Open Authorization** (**OAuth**) to enable authentication via third-party providers such as GitHub, Facebook, Twitter, Google, and many others.

Now that we have understood what Strapi is all about, let's see how we can prepare our development environment to start developing our API.

Preparing the development environment

Before we start developing our API, we will need to prepare our development environment first. Next, we will look at the components and packages that we will be needing throughout the book.

> **Note**
>
> You can install the packages in any order you want. Some of the packages here are not a must-have, but it's highly recommended to install them to avoid having any issues while following the examples in this book.

Installing Node.js

To install Node.js, head to the official website, `https://nodejs.org`, and download the **long-term support** (**LTS**) version that matches your operating system. At the time of writing this book, version 16.13.2 is the latest LTS version and the version recommended by Strapi, as illustrated in *Figure 1.2*. Only LTS versions are supported by Strapi, the other versions of Node.js are not guaranteed to be compatible.

Node.js® is a JavaScript runtime built on Chrome's V8 JavaScript engine.

New security releases now available for Node.js 12, 14, 16, and 17 release lines

Download for macOS (x64)

Or have a look at the Long Term Support (LTS) schedule

Figure 1.2: Node.js LTS version

Once you have installed Node.js, open your favorite terminal and run the following command to verify the Node.js version:

```
node -v
```

You should see the version of the installed Node.js on the terminal.

Installing Visual Studio Code (optional)

Visual Studio Code (**VS Code**) is a feature-rich and powerful **integrated development environment** (**IDE**), and we will be using it as our default editor. You are free to use whichever editor you feel comfortable with. However, we highly recommend installing and using VS Code to follow along with the examples in this book.

To download and install VS Code, head to `https://code.visualstudio.com` and download and install the appropriate build for your operating system.

Installing Yarn

Yarn (`https://yarnpkg.com`) is a JavaScript package manager. We will be using it as our default package manager since it's the package manager used by Strapi itself.

To install Yarn, open your favorite terminal and execute the following command:

```
npm install -g yarn
```

Once the installation is complete, run the following command to verify that Yarn has been installed successfully:

```
yarn -v
```

The version of Yarn should be displayed in the terminal. At the time of writing this book, version 1.22 is the latest version.

Installing Docker (optional)

We will be using Docker to install and manage a Postgres database for our API in a later chapter of the book. Docker can help us easily work with different database systems such as Postgres, MySQL, or MongoDB. To install Docker, head to `https://docker.com/get-started` and download and install Docker Desktop for your operating system.

Once you have installed Docker, execute the following command in your terminal to verify the installation:

```
docker -version
```

The Docker version should be displayed in the terminal. At the time of writing this book, version 20.10 is the latest version.

Alternatively, you can install Postgres using a different method of your choice.

Installing Postman

Postman is a great API client, and we will use it to interact with our API. Head to the Postman website at `https://www.postman.com/downloads/` and download the version matching your operating system.

Once we have the development environment set up, we can proceed with creating our API.

Creating a Strapi application

Begin by running the following command on your terminal:

```
yarn create strapi-app strapi-lms --quickstart
```

The preceding command will set up a project using the latest version of Strapi with the default settings. Using a SQLite database, start the server on port `1337` and launch the admin dashboard.

> **Note**
>
> If we remove the `--quickstart` flag, we will enter manual setup mode, where we will be asked a few questions to configure Strapi.
>
> If the admin panel did not launch automatically, you can open your browser and navigate to `http://localhost:1337/admin`.

The first time you log in to the admin dashboard, you will be presented with a form to create the first administrator user. Complete the form to create an administrator user and sign in to the admin panel.

Overview of the admin panel

We will discuss the admin panel in greater detail in *Chapter 4*, *An Overview of the Strapi Admin Panel*. However, for now, we will just give a quick overview of the admin panel layout.

The admin panel is easy to navigate. On the left-hand side, we have our main control sidebar. It can be divided into three main categories, as follows:

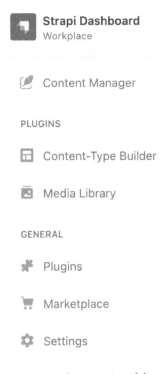

Figure 1.3: Strapi main sidebar

Content Manager

This is where the API content lives, and you can manage API content from this section. At the moment, there is only one model, **Users**. We will see more models here as we progress in developing the API.

Plugins

The **PLUGINS** section allows you to customize Strapi. There are two essential plugins available out of the box: the **Content-Type Builder** and **Media Library** plugins.

The **Content-Types Builder** plugin is the core of Strapi customization; we will use it to create new models in our API and create a relationship between those models. As this plugin is critical in developing the API, we have dedicated *Chapter 3*, *Strapi Content-Types*, to discussing it in greater detail.

The **Media Library** plugin, as the name suggests, is used to manage all API media files.

General

This section contains Strapi settings where you can use the **Marketplace** to install new plugins, configure plugin settings, and edit general settings such as adding additional administrator users.

The `--quickstart` flag will bootstrap the Strapi application and start the development server for us. While you are developing a Strapi application, you will want to start and stop the server yourself. Let's see in the next section how to work with Strapi scripts to manage the server.

Understanding server scripts

Strapi comes with a few scripts that can be used in managing the development server as well as starting the server in production. If you open the `package.json` file, you will see the following four scripts:

```json
"scripts": {
  "develop": "strapi develop",
  "start": "strapi start",
  "build": "strapi build",
  "strapi": "strapi"
},
```

Figure 1.4: Strapi server scripts

The develop script

The `develop` script will start the server in development mode, with autoreload enabled. Basically, it will watch for any changes in the project files and restart the server if there are any. This script is intended for local development, and it should never be used in a production environment.

The start script

The `start` script will start the server with autoreload disabled. This script is intended to start the server in a production environment.

The build script

This script allows you to rebuild the admin panel. The Strapi admin panel is built using React.js, but in some situations, you might want to customize or extend the admin panel. In such cases, you will need to rebuild the admin panel again using the `build` script. The admin panel is built once when you have created a project and every time you install a plugin that requires changes to the admin panel.

The Strapi script

This script is an alias to the Strapi **command-line interface** (**CLI**), and we can use it to generate new content in our system. We will explore the Strapi CLI in the next chapter.

Summary

In this chapter, we started by explaining the concept of a headless CMS and saw how it is different from a traditional CMS. We then introduced Strapi, an open-source headless CMS, and listed the benefits of using Strapi in developing APIs.

Then, we started preparing the development environment and installed the requisite packages and software, as well as creating a sample application that we will be using throughout the rest of this book. Finally, we had a quick look at the different server scripts offered by Strapi.

In the next chapter, we will have a deeper look at the typical structure of a Strapi application. We will also build our first API in the system and learn how to use the Strapi **Content-Types Builder** plugin to define fields needed for the API.

2
Building Our First API

In this chapter, we will create our first API endpoint. We will explore the structure of a **Strapi** API project, and what modules and files Strapi creates for us. We will touch on the concept of a **content-type** and explore the structure and commands of a typical Strapi application.

The topics we will cover in this chapter are as follows:

- Exploring the project structure of a Strapi application
- Defining the requirements – what are we going to build?
- Creating our first content-type using the admin panel and the **command-line interface (CLI)**
- Clarifying Strapi terminology – content-type, resources, and object types
- Interacting with our API
- Understanding the makeup of an API – routes, controllers, services, and models

Exploring the project structure of a Strapi application

When we ran the `strapi-app` CLI tool with `yarn create strapi-app`, the command generated a new blank project for us. Let's spend some time exploring the structure of the project and the packages that were installed for us.

The commands to run Strapi

Out of the box, the `package.json` file includes four Strapi-related commands that provide aliases for the Strapi CLI:

- **Development mode**: While developing, we will run the API using the `yarn develop` command. This is an alias for `strapi develop`, which runs the Strapi instance in development mode, enabling features such as auto-reload and writing files to the code base when we change options in the admin panel. It also builds the admin panel.

- **Production mode**: For running in production, we can use `yarn start` (which is the alias for `strapi start`). This disables auto-reload and writing files into the code base.

- **Build the admin panel**: The final command is `yarn build` (an alias for `strapi build`), which builds the admin panel. This is useful for customizing the admin panel, which we will do in *Chapter 4, An Overview of the Strapi Admin Panel*.

- **The Strapi CLI alias**: This is a handy alias to run the Strapi CLI commands. The CLI provides the previous three commands (`develop`, `build`, and `start`) and others to generate the APIs, reset the admin user, and so on. You can get a list of the available commands and their documentation by running `yarn strapi -help`.

The project structure

The project structure is simple:

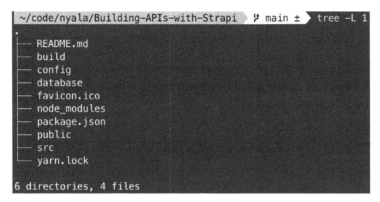

Figure 2.1: The project structure of a Strapi project

Let's go over the preceding figure in detail:

- `config/`: This contains various application configurations, such as `database.js`, where we can define our database configurations and settings; `api.js`, where we can define settings related to how to query the API; and `admin.js`, where we can configure settings related to the admin panel. Other than that, `server.js` contains a couple of options that we may update, such as `PORT` and `HOST`, where the API will run any other configurations related to the server.

> **Overriding the Config Folder for an Environment**
>
> Strapi also provides the ability to override the configurations in the `config` folder for a specific environment.
>
> So, for example, we can create a server configuration file for the `production` environment by creating a file called `./config/env/production/server.js`. Strapi then checks the NODE_ENV variable, and if it is set to `production` (which matches with the folder name we created), then the production configuration will be merged with the base configuration, overriding any setting. This means, for example, that we can use a database such as SQLite in *development* and use a PostgreSQL database in *production*.

- `public/`: This contains static files such as images that we want to make accessible through our API.

- `build/`: This contains the build files for the admin panel. This file is ignored in `.gitignore` and gets automatically generated when the admin panel is rebuilt.

- `src/admin/`: This contains files related to customizing the Strapi admin panel.

- `src/extensions/`: We will use this folder to extend and configure the plugins provided by Strapi. We will do this in *Chapter 8, Using and Building Plugins.*

- `src/api/`: This is where we will spend most of our time. This is the folder where our *content-types* will be defined, and the code for them is auto-generated by Strapi.

There are other folders (and files) that we can create as well, for example, to add **middleware** and **hooks,** or customize API responses globally. We will visit these throughout the book when we need to use them.

Under the hood – the components that make up Strapi

Let's dig a bit deeper into the dependencies that Strapi installs. These dependencies are implementation details of Strapi. While there is no need to understand them to use Strapi and build APIs, they're still interesting to know about at a high level to allow a deeper understanding of Strapi, its architecture, and its philosophy:

```
"devDependencies": {},
"dependencies": {
  "@strapi/strapi": "4.0.0",
  "@strapi/plugin-users-permissions": "4.0.0",
  "@strapi/plugin-i18n": "4.0.0",
  "sqlite3": "5.0.2"
},
```

Figure 2.2: The packages used by a Strapi project

Let's look at this figure in detail:

- `@strapi/strapi`: This is the core package for Strapi; it provides the HTTP layer that our API will sit on. It's also responsible for managing core functionality, such as bootstrapping the initial setup of the database and admin users, and exposing and loading services, hooks, extensions, plugins, and all the components that form Strapi's architecture. It's the glue for everything that happens in Strapi. It's built on top of **Koa**, a popular node web framework.

- `@strapi/plugin-user-permissions`: Almost everything in Strapi is a plugin; this one includes capabilities for creating users and managing permissions. This is one of the core principles of the philosophy behind Strapi's architecture. Strapi's core is intentionally very slim; the rest of it – including functionality that we might think of as core, such as creating content-types and endpoints – is implemented in the form of plugins. This opens the door to an infinite number of possibilities and capabilities. We will discuss users and roles in greater detail in *Chapter 7, Authentication and Authorization in Strapi.*

- `@strapi/plugin-i18n`: This plugin allows a user to create localized content in different languages.

- `sqlite3`: This package allows us to use SQLite as our database. It was included in the project because we specified the `--quickstart` flag when creating the project.

Next, we will talk about the project that we will be building throughout this book and the first API we will create for this project.

Defining the API requirements

We will be building an API for a **Learning Management System** (**LMS**). This is a system for an educational institution that runs *classrooms* for different *courses*. *Students* can enroll in these classrooms, and *teachers* can create new *tutorials* and assign them to classrooms. Another (non-core) functionality might include creating *quizzes* that the students can answer and the teachers can mark. An *admin* can also access the system to add new students, teachers, and classrooms, and manage all entities:

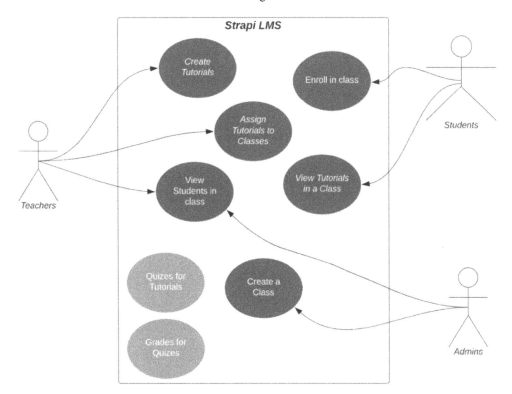

Figure 2.3: A use case diagram for the main actors and functionalities of the LMS

In the next section, we will create our first API, which we will use to manage a *classroom*. The endpoint will allow us to perform **CRUD** operations; that is, we will be able to **Create**, **Read**, **Update**, and **Delete** classrooms.

Creating our first content-type – a classroom

We will create our first content-type, which will represent the `classroom` entity. For now, a classroom will have `name`, `description`, and `maximum students` values. Let's get started:

1. Run the Strapi admin panel, normally at `http://localhost:1337/admin`.

> **Remember**
>
> By default, Strapi uses port `1337`. This value is used in the `server.js` file and can be overridden using the `PORT` environment variable. We will discuss this later on in *Chapter 9, Production-Ready Applications*.

2. Navigate to **Content-Type Builder**, under the **PLUGINS** section.

3. Click on **Create new collection type**.

4. For **Display name**, enter `Classroom`; leave the other fields as they are.

> **Important Note**
>
> **API ID (Singular)** and **API ID (Plural)** are auto-populated from the **Display name** field. **API ID (Singular)** is used in generating the database tables while **API ID (Plural)** is used to access the content-type via the API URL; we will see this in action later on in this chapter.

5. Then, click on **Continue**:

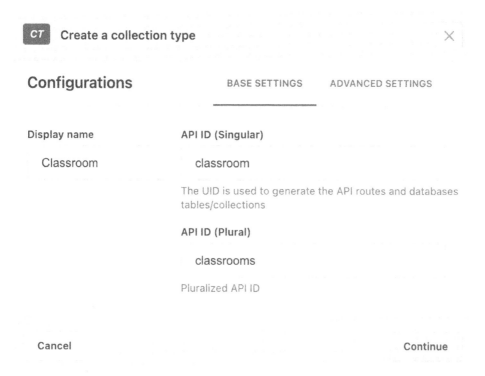

Figure 2.4: The Create a collection type form

6.   Select **Text**, and then choose **Short text** in the **Type** field and enter name in the **Name** field:

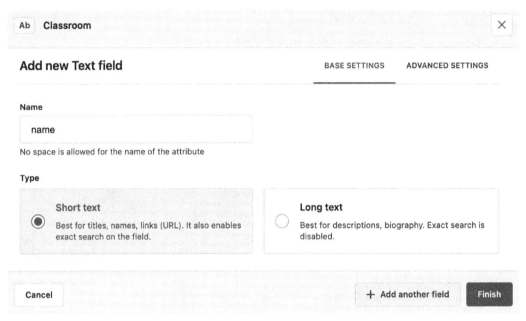

Figure 2.5: The Add new Text field form

7.   Then, click on **+ Add another field** and repeat the same for the other two fields:

a. **Name**: description

Type: **Long text**

b. **Name**: maxStudents

Type: **Number**

Number format: **Integer**

8.   Then, click **Finish**.

9.   Finally, click **Save**.

This will create the content-type with the fields that we specified and also restarts our development server. If all went well, then you should see **Classroom** under **Content Manager** | **COLLECTION TYPES** on the right:

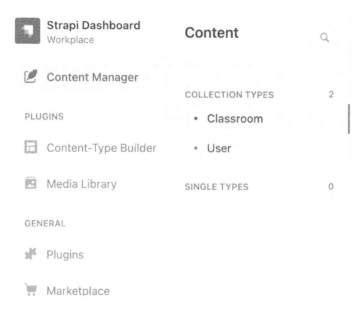

Figure 2.6: The newly added collection type in the navigation menu

Next, let's create a classroom data entry.

10. Click on **Classrooms**, and then click **Add new entry** to add your first classroom entry:

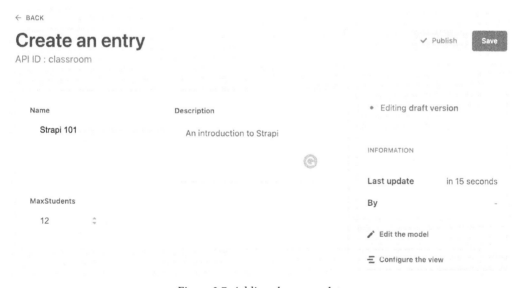

Figure 2.7: Adding classroom data

11. Click **Save** to save the contents, and then click **Publish** to change the entry type from draft to published. You should see that the **Publish** button has changed to **Unpublish** and a note saying **Editing published version** has appeared, showing that we are now working on the published version:

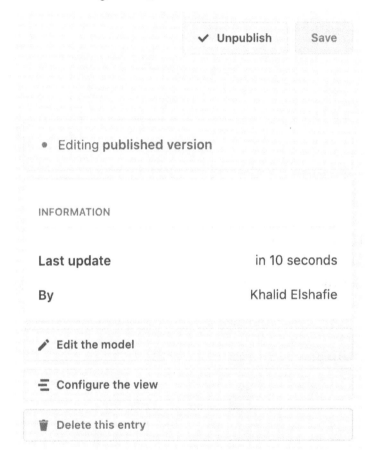

Figure 2.8: The Strapi save and publish dialog

Using the Strapi CLI instead of the admin panel

The admin panel is very user-friendly, but Strapi also gives us the option to perform any action through the CLI as well.

The generate CLI command can be used to generate *API*, *controllers*, *content-types*, and *plugins*, among others. The generate command is fully interactive; to use this command, run the following CLI command and then follow the onscreen instructions:

```
yarn strapi generate
```

> **Important Note**
>
> If you have already created the `classroom` content-type using the admin panel, then you will not be able to create it using the CLI `generate` command since the content-type already exists. Delete the existing `classroom` content-type created with the admin panel before using the interactive generate `CLI` command.

We have already used the Strapi CLI to scaffold the project skeleton, and we have now used it to create an API. The Strapi CLI is very powerful, and it provides all kinds of functionalities to make the developer experience better. We will not go over the details of these commands, as they are well covered in the documentation, but we advise you to (at least) run the CLI help with `yarn strapi -help` to gain an idea of what can be accomplished with the CLI. You can run `yarn strapi help` followed by the command name to get information about any of the commands. In practice, this is what it looks like:

yarn strapi help generate

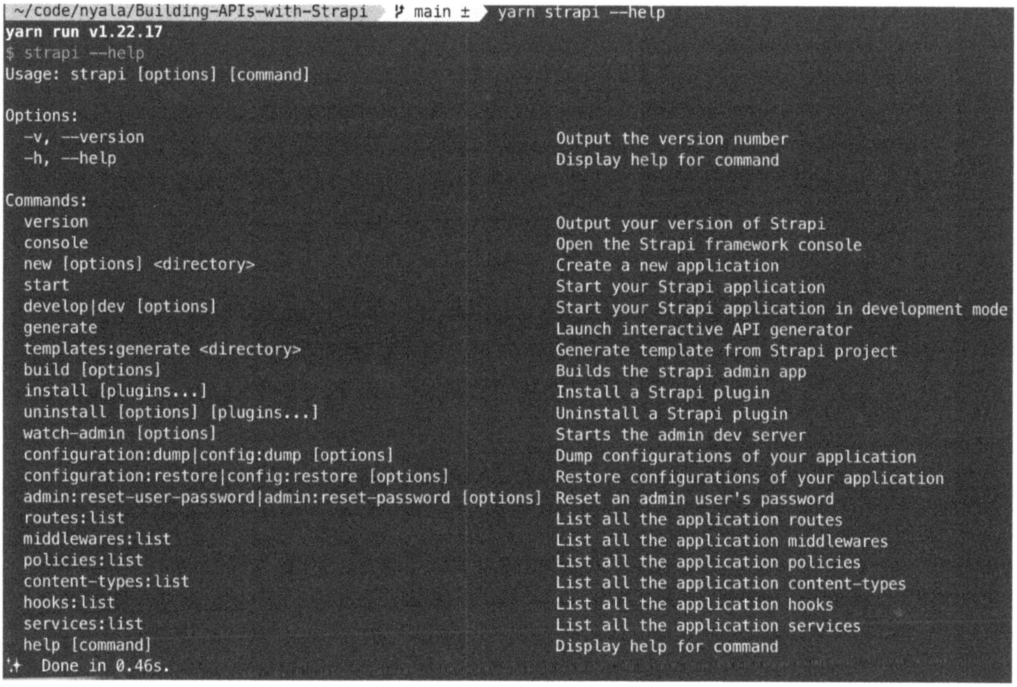

```
~/code/nyala/Building-APIs-with-Strapi   ⎇ main ±   yarn strapi --help
yarn run v1.22.17
$ strapi --help
Usage: strapi [options] [command]

Options:
  -v, --version                                          Output the version number
  -h, --help                                             Display help for command

Commands:
  version                                                Output your version of Strapi
  console                                                Open the Strapi framework console
  new [options] <directory>                              Create a new application
  start                                                  Start your Strapi application
  develop|dev [options]                                  Start your Strapi application in development mode
  generate                                               Launch interactive API generator
  templates:generate <directory>                         Generate template from Strapi project
  build [options]                                        Builds the strapi admin app
  install [plugins...]                                   Install a Strapi plugin
  uninstall [options] [plugins...]                       Uninstall a Strapi plugin
  watch-admin [options]                                  Starts the admin dev server
  configuration:dump|config:dump [options]               Dump configurations of your application
  configuration:restore|config:restore [options]         Restore configurations of your application
  admin:reset-user-password|admin:reset-password [options] Reset an admin user's password
  routes:list                                            List all the application routes
  middlewares:list                                       List all the application middlewares
  policies:list                                          List all the application policies
  content-types:list                                     List all the application content-types
  hooks:list                                             List all the application hooks
  services:list                                          List all the application services
  help [command]                                         Display help for command
✨  Done in 0.46s.
```

Figure 2.9: The commands available through the Strapi CLI

Before we start testing and interacting with our API, we will briefly cover some of the terminologies we've used in Strapi so far.

Clarifying Strapi terminology

One of the goals of Strapi is to build APIs easily and quickly without being bogged down in the implementation details of specific architectural styles (such as **GraphQL** or **REST**) or implementation details (such as which database and database connectors to use). To achieve this technological agnosticism, Strapi uses terminology that sits *on top* and abstracts such details. We will quickly cover this terminology so that we use the same vocabulary that Strapi uses throughout the book and will also be able to relate it to concepts we might already know from our previous exposure to other API frameworks.

Content-types, resources, object types, and models

Content-types are the fundamental building blocks in Strapi. If you have built REST APIs before using a more low-level framework, you might think of the content-types as *resources*. When building an API in the REST architectural style, we develop our endpoints and our interactions around resources. They are the fundamental building blocks. They are a conceptual mapping to an entity, rather than the entity itself (think of *classes* and *objects* in object-oriented programming):

> *"The key abstraction of information in REST is a resource. Any information that can be named can be a resource: a document or image, a temporal service (e.g. 'today's weather in Los Angeles'), a collection of other resources, a non-virtual object (e.g. a person), and so on... A resource is a conceptual mapping to a set of entities, not the entity that corresponds to the mapping at any particular point in time."*
>
> *– Taken from Roy Fielding's dissertation on the REST architectural style*

If you are coming from a GraphQL background, then a similar concept also exists in *object types*. These are, again, a representation of an abstract concept.

In the Strapi world, these resources (or object types) are called content-types. A content-type represents an abstract concept. When we create a content-type called `classroom`, we are representing the abstract concept of a classroom in our educational institution, regardless of whether we will expose that concept with a REST or GraphQL API.

Models are how Strapi defines these content-types in code. They are a database-agnostic representation of the content-type's fields, the types of these fields (that is, a number or text field), and any constraints applied to them, as well as the relationships between the content-types. This representation is eventually mapped to a real database entity.

Content-types are Strapi's API-agnostic way to define and think about the entities that make up a system; models are Strapi's database-agnostic way of representing content-types in code. We will dive deeper into these concepts in *Chapter 5, Customizing Our API*.

In Strapi, there are three content-types – *collection types*, *single types*, and *components*. We will discuss these in detail in the next chapter, but for now, it's sufficient to know that our classroom is a collection type because we will have multiple classrooms running in our school.

From this point onward, we will use *content-type* (not *resource*) to be consistent with the Strapi documentation and the admin panel.

Interacting with the Strapi API

To access any API endpoint in Strapi, we will use the HOST server and PORT, followed by /api/, and then the **API ID (Plural)** value.

> **Remember**
>
> The **API ID (Plural)** value was generated when we created the classroom content-type earlier in this chapter. It is usually the plural form of the content-type name.

So, to access the classroom endpoint on our API, we can use the following URL: http://localhost:1337/api/classrooms. This can be done from the browser since it's a GET request, but let's get into the habit of using **Postman** early on. The expectation is that this endpoint should return a list of classrooms:

Figure 2.10: Calling the API endpoint using Postman

Setting permissions

We received a 403 Forbidden HTTP status code from the backend. Strapi secures API endpoints by default, so we will have to explicitly tell it to allow access to the API for all users. To do that, let's head back to the admin panel:

1. Click **Settings** and then click **Roles** (under **USERS & PERMISSIONS PLUGIN**). We can see that Strapi has already created two roles for us, Authenticated and Public.

2. Click **Public**, and then under **Permissions**, expand the **Classroom** panel and check the **find** checkbox to select the **find** action. By doing this, we are instructing Strapi to allow unauthenticated users (that is, Public users) to perform this action. Note also that the route associated with the action, /api/classrooms, is highlighted on the right of the pane:

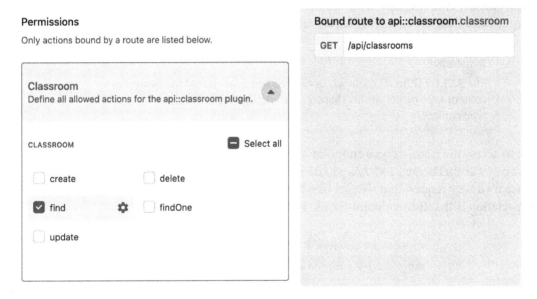

Figure 2.11: Updating the classroom content-type permissions to allow a find action

3. Click on **Save**, and then head back to Postman and try the request again. Voila! Now our API returns a 200 OK HTTP status code. The response body contains a data array that holds the result of the API call, and a meta object that contains metadata, such as the pagination information. We will discuss this in *Chapter 6, Dealing with Content*:

Figure 2.12: An API call to get all classrooms data

4. Now, let's go back to the **Roles** screen (under **Settings**) and allow the `Public` role to perform the create action by selecting the **create** checkbox:

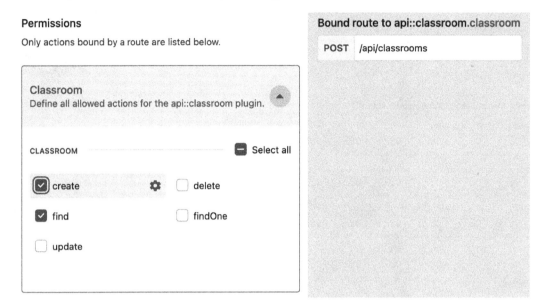

Figure 2.13: Updating the classroom content-type permissions to allow a create action

5. Next, let's perform a `POST /api/classrooms` request to create a new classroom. When sending a `POST` or `PUT` request, we will need to wrap the request payload in a `data` object, the properties of which are the content-type we want to create or update. In this case, the data object will contain the `classroom` entity to be added. Make sure that the request content-type is set to `JSON` in Postman:

Figure 2.14: Creating a classroom entity via the API endpoint using Postman

As you can see, we get a 200 OK HTTP status code as a response and the body of the classroom content-type we just added, including fields that were returned by Strapi after saving to the database (for example, the id, published_at, created_at, and updated_at timestamps).

If we go back and call GET /api/classrooms, then we will get back two classrooms instead of one – the Strapi 101 classroom that we added from the admin panel and the one we just added from Postman.

What did we just create?

We just created our first content-type representing the *concept of a classroom*. Strapi created a REST API with five endpoints for us for the various CRUD operations – `create`, `read` (a single entity or all entities), `update`, and `delete`:

- `GET /api/classrooms`: To list all classrooms
- `GET /api/classrooms/:id`: To get a single classroom using its ID
- `POST /api/classrooms`: To create a new classroom
- `PUT /api/classrooms/:id`: To edit a classroom with a matching ID
- `DELETE /api/classrooms/:id`: To delete a classroom with a matching ID

Now, you can go back to **Users & Permissions** and allow the `Public` role to perform all five actions; then, you can test them in Postman.

If you have ever created a REST API in a lower-level framework such as **Express** or Koa, you must appreciate how much time Strapi has saved us. Within a few minutes, we have created a REST API that is ready to use.

Strapi did many things under the hood for us with just a few clicks:

- It created five routes for our content-type that follow the `REST` naming conventions.
- It allowed us to define the fields that are part of that content-type.
- It gave us a simple publishing system where we can define an entity as `published` or `draft`.
- It created the tables, columns, and database entities required to support this content-type.
- It implemented sensible defaults for us, such as adding a unique ID auto-increment field to our content-type and adding update and creation timestamps to our entities.
- It even took care of pluralizing the endpoint name (for example, `/classrooms` and not `/classroom`) and the name of the database table created.

Instead of sorting out all of these details ourselves (think of authentication, database layers, and route definitions), we are left to concentrate on what really matters – the application that we're building. Some experienced developers might be skeptical at this point and think of *the 80–20 paradox*. A lot of frameworks make it straightforward to implement the easy scenarios (that is, the ones that make up 80% of use cases), but then they're a nightmare when we want to build something unique (that is, 20% of use cases).

In the next section, we will explore in more detail what Strapi has done for us so far. We will look into some of the design and architecture decisions that make Strapi extremely convenient for tackling this 80% of use cases, but at the same time, make sure that it never limits us as developers when we need to do something different.

> **What about GraphQL?**
>
> Out of the box, Strapi gave us the REST endpoints around the content-type we created. Supporting GraphQL is as easy as installing a plugin. Once we do this, we're ready to expose our content-types in a GraphQL API as well as a REST API. We will cover the Strapi support for GraphQL in detail in *Chapter 10, Deploying Strapi*.

Understanding the makeup of a Strapi API

Several files were created for us under the `src/api` folder when we created the `classroom` content-type. Now, we will take a look at the updated project structure and consider a general overview of what was created.

Routes, controllers, services, and models

When we created our first content-type in the admin panel, several files were created under the `src/api` folder:

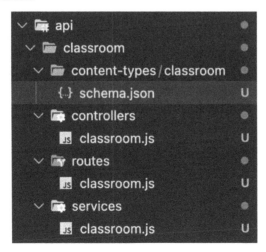

Figure 2.15: The classroom content-type folder structure

Let us have a quick overview of the route components of a Strapi API, that is the routes, controllers, services and models:

- There are two types of routes in Strapi, a core route and a custom route. Core routes are automatically created by Strapi when a new content-type is created; the core routes are `find`, `findOne`, `create`, `update`, and `delete`. The `routes/ {content-type}.js` file contains the core *route* (or endpoint) definitions. This file uses a `createCoreRouter` factory function that automatically generates the core routes. Each one of these routes maps to a *controller* action. We will discuss customizing the routes in, *Customizing Our API*.

- Controllers live under `controllers/{content-type}.js` (for example, `controllers/classroom.js`). Controllers are responsible for handling incoming requests and each controller defines a set of methods called *actions*. Each action is mapped to a specific route. The file initially contains a single `createCoreController` factory function that automatically generates the core actions and allows us to build custom ones.

- Controllers use *services*. Services live under `services/{content-type}. js` (for example, `services/classroom.js`). Services are helpers used by the controllers to perform various operations, including CRUD operations but also custom services for other supporting operations, such as sending an email and sending notifications.

- The default services for handling CRUD operations provided in the core services from Strapi deal with *models*. Models are the code representation of the *content-types* we created in the admin panel in the previous section. They map to database entities (for example, *tables* and *columns* in a relational database such as PostgreSQL). The definition for these models lives under `content-types/{content-type}/schema.json` (for example, `content-types/ classroom/schema.json`). This file includes the field names (attributes), types, and other information related to the content-type.

We will explore content-type and models more thoroughly in *Chapter 3, Strapi Content-Types*:

```
{..} schema.json U ✕

src > api > classroom > content-types > classroom > {..} schema.json > ...
  1    {
  2      "kind": "collectionType",
  3      "collectionName": "classrooms",
  4      "info": {
  5        "singularName": "classroom",
  6        "pluralName": "classrooms",
  7        "displayName": "Classroom"
  8      },
  9      "options": {
 10        "draftAndPublish": true
 11      },
 12      "pluginOptions": {},
 13      "attributes": {
 14        "name": {
 15          "type": "string"
 16        },
 17        "description": {
 18          "type": "text"
 19        },
 20        "maxStudents": {
 21          "type": "integer"
 22        }
 23      }
 24    }
 25
```

Figure 2.16: The model that was created for the classroom content-type

There is one more module that is not created by default but worth mentioning now, which is the `content-types/{content-type}/lifecycles.js` file. This file allows us to manage the model life cycle. The life cycle is a hook that is used to run custom code at different parts of the life cycle of an entity – for example, before it's created, after it's created, when it's updated, and when it's deleted.

We will talk in more detail about customizing the routes, controller, and life cycles in *Chapter 5, Customizing Our API*:

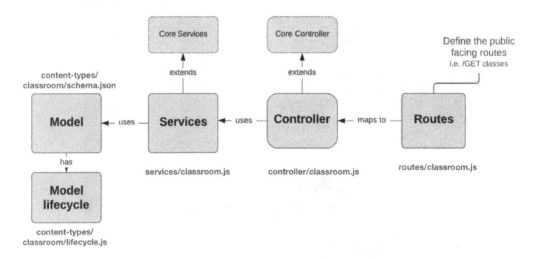

Figure 2.17: An overview of the Strapi core components

Why Are All These Files Empty?

Other than the `content-types/classroom/schema.json file`, all the files that were created are empty. Yet, we've tested that our REST API works when we queried the list of classrooms, and we even created one through the API. This is because controllers inherit their behavior from the core components provided by Strapi, which implement default behavior for the main CRUD operations. The files are there in case we want to override one (or all) of these components. Almost always, the defaults provided are enough, but when we need to implement custom behavior, all we need to do is override these defaults.

Content-types in code

Whether we defined our content-type through the admin panel or the CLI, these definitions end up in code under `src/api`. This is one of the clever design decisions behind Strapi; having these defined in code – as opposed to, for example, having them defined in a configuration saved in the database – makes them very easy to move between environments (for example, from your local development environment to your testing environment, and eventually your production environment). It also makes it easy to track the progress of our content-type definitions because they live in the source code and, therefore, they will be governed by living inside a source control system.

> **Not Everything We Did Lives in Code**
>
> You might have noticed that the permissions we applied to the content-type (for example, allowing unauthenticated users to access the `/api/ classrooms` endpoints) don't live in code. This means that if we deploy our code in a different environment, or if we drop the database and recreate it, we have to manually reapply the permissions. This quickly becomes inconvenient when managing multiple environments, and we will learn about practices (and plugins) to handle these situations in *Chapter 11, Testing the Strapi API.*

Summary

In this chapter, we built our first REST API to manage our new content-type – `classroom`. We defined the content-type for a classroom, defining the fields that make up that type, and Strapi built an API for us where we can create new classroom entities, retrieve the list of classrooms, update existing classrooms, and delete classrooms (in other words, perform *CRUD* operations). We saw how Strapi allowed us to create these endpoints without any coding, saving us potentially hours of development time. We also touched on the permissions model of Strapi, which allows us to interact with these endpoints.

We also dug a bit deeper into what Strapi and the Strapi CLI create for us when we create a new project or add a new content-type. While Strapi makes it extremely easy to create APIs without worrying about specific details, knowing a little bit about what goes on under the hood helps us appreciate our design options and get a feel for how it is architected. Understanding this early on in this process can come in useful later on when considering things such as extensibility.

In the next chapter, we will explore more fundamental concepts relating to *content-types* in Strapi. As this is the core concept of Strapi, a strong grasp of it will help us as we tackle more complex scenarios.

3
Strapi Content-Types

In the previous chapter, we briefly introduced the **Content-Type Builder** plugin. In this chapter, we will discuss the **Content-Type Builder** plugin in greater detail, and we will see how it is used in building and developing the **API** (short for **application programming interface**) as well as creating relationships between different entities in our API.

These are the topics we will cover in this chapter:

- What is the **Content-Type Builder** plugin?
- Creating and managing content-types
- Understanding relations in Strapi
- Interacting with the API endpoints from Postman
- Differences between **COLLECTION TYPES** and **SINGLE TYPES**

What is the Content-Type Builder plugin?

As we briefly discussed in *Chapter 2, Building Our First API*, content-types are Strapi's API-agnostic way to define the entities that make up a system. The **Content-Type Builder** plugin is what we use to create and manage those content-types. It is one of Strapi's core plugins, it comes installed and enabled with Strapi by default, and it cannot be deleted.

> **Note**
>
> In production environments, the **Content-Type Builder** plugin is read-only and cannot be used to alter and change the API content-types.

Content-types in Strapi are categorized into three categories—**COLLECTION TYPES**, **SINGLE TYPES**, and **COMPONENTS**. The differences between them are outlined here:

- **COLLECTION TYPES** are content-types that can manage several entries. For example, we have multiple classrooms in our system, thus we used the collection types to create the **Classroom** content-type in the previous chapter.

- **SINGLE TYPES**, as the name suggests, are single-entry content-types that you can visualize as a single-row table. We can use them to manage things such as application settings.

- **COMPONENTS** are a custom data structure that can be used in **COLLECTION TYPES** as well as **SINGLE TYPES**. For example, we can create a **Full Name** component that is comprised of two text fields and then use this component in the **COLLECTION TYPES** or **SINGLE TYPES** content-types. **COMPONENTS** are of value if you are using the Strapi admin panel to manage the API contents.

You can see an overview of the different content-type categories in the following screenshot:

Figure 3.1: Strapi content-type categories

Now that we understand what the **Content-Type Builder** plugin is all about, let's use it in creating the next content-type in our API.

Creating and managing content-types

Let's create a **Tutorial** content-type. Since we will have multiple tutorials in our system, we will use the **COLLECTION TYPES** category to create a **Tutorial** content-type.

From the **Content-Type Builder** plugin page, click **Create new collection type**. The **Create a collection type** modal will appear. It has two tabs—the **BASIC SETTINGS** and **ADVANCED SETTINGS** tabs.

In the basic settings tab, we have three fields—the **Display name** and **API ID (Singular)** and **API ID (Plural)** fields. These are described in more detail here:

- The **Display name** value is what we want to call our content-type, so enter the value `Tutorial` in this field.

- The **API ID** fields are automatically generated from the Display name, the Singular format is used in creating the database table as well as generating the API routes, while the Plural format is used in the API URL. For example, the Tutorial content-types can be accessed using the URL `/api/tutorials`.

> **Best Practice for Naming Content-Types**
>
> It is recommended to use the singular form when creating new content-types as Strapi will use the plural form to create a database table and the resources' API URLs.

The **ADVANCED SETTINGS** tab allows us to configure additional properties for the content-type we are about to create, as illustrated in the following screenshot:

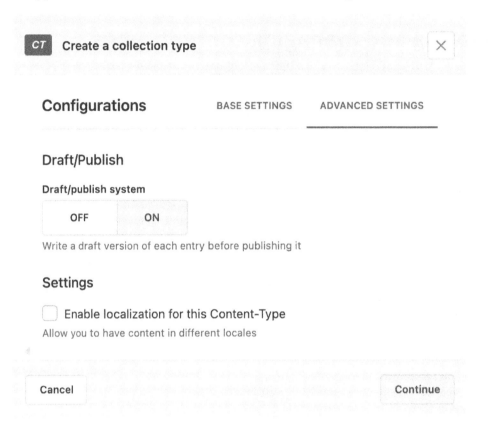

Figure 3.2: Collection types advanced settings

There are two options available in the advanced settings. We are not going to modify any of them now, but it is worth exploring and understanding what the available options are. Let's have a look at them here:

- **Draft/publish system**: This option allows us to turn **ON/OFF** the publishing feature for the content-types. When this feature is turned on, all contents are saved as a draft by default when using the Strapi admin panel and it requires publishing to be available in the API response. The default value for this option is **ON**.

- **Enable localization for this Content-Type**: Checking this box allows to you manage the content-type in various locales.

We will leave all the options at their default values. Click the **Continue** button to continue creating a content-type.

> **Note**
>
> At this stage, the new **Tutorial** content-type has not yet been created. A new content-type is considered created only once it has been saved.

Once we click the **Continue** button, we will be presented with the content-type fields creation page, as illustrated in the following screenshot:

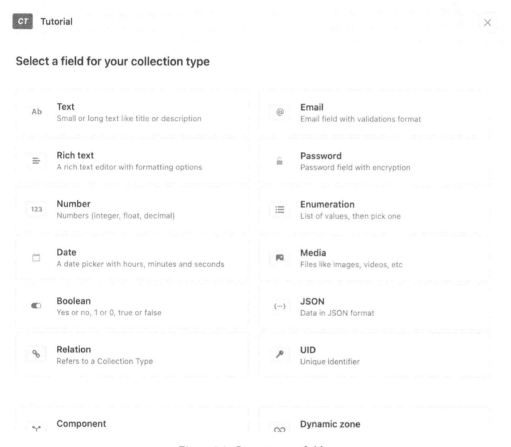

Figure 3.3: Content-type fields

Here is a list of the fields that we want to have in our **Tutorial** content-type:

- **title**: Each tutorial must have a title.
- **slug**: A **Uniform Resource Locator** (**URL**)-friendly field that must be unique and is driven from the title.
- **type**: A tutorial can be either a **video** or **text** tutorial.

- **url**: If the tutorial type is **video**, then this field will hold a link to the tutorial video.

- **contents**: If the tutorial type is **text**, then this field will hold the contents of the tutorial.

- **classroom**: Relationship to the classroom. A tutorial belongs to a single classroom.

Creating a title field

Let's start by creating a title field first—we will choose **Text** for the field type. Once we have selected the field type, a new pop-up window will be shown with a similar layout to the collection types creation popup we saw earlier.

In the field configuration screen, enter the value `title` as the field name and then select the **ADVANCED SETTINGS** tab to switch to the advanced settings. In this setting screen, we can configure additional properties to the field we are about to create; those settings will vary slightly depending on the field type. However, here are the common settings available to all of the field types:

- **Default value**: Allows us to set a default value for the field

- **RegExp pattern**: Allows us to set a **regular expression (regex)** that can be used in validations

- **Required field**: Indicates whether a field is a required field or not

- **Unique field**: Indicates that the value should be unique within the column in the database

- **Maximum length/Minimum length**: Allows us to set the minimum or maximum length for the field

- **Private field**: Allows us to make the field private and not show it in the API response

You can see an overview of these settings in the following screenshot:

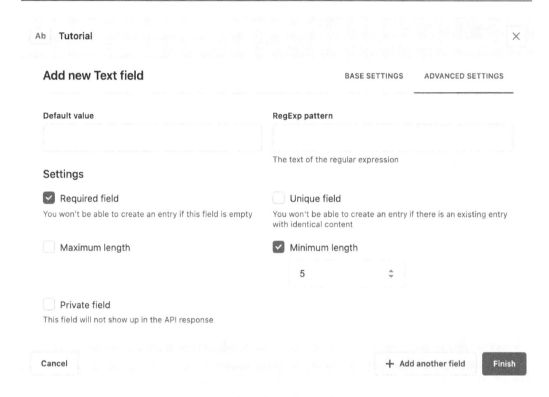

Figure 3.4: Text field advanced settings

Going back to the title field settings, let's mark this field as required and set a minimum length of 5 characters for the title and then click the + **Add another field** button.

Creating a slug field

The next field we want to create is a slug field; we will also choose **Text** for the field type. Proceed as follows:

1. Select the **Text** field type.

2. Enter the value slug as the name of this field.

3. Switch to the **ADVANCED SETTINGS** tab.

4. Check the **Required field** and **Unique field** boxes.

5. Click + **Add another field** to continue.

> **What about the UID Field for the Slug?**
>
> You might have noticed a field type called **UID**. The **UID** field allows you to create text without spaces, and it must be unique in the database. You can also specify another field as the drive for the UID value. However, this option is only practical if you are using the Strapi admin panel to manage the content, and it will not work with an API client such as Postman. The reason for this is the fact that the logic behind creating the field is in the admin panel code itself and not in the underlying API. In *Chapter 5, Customizing Our API*, we will see how to use Strapi life cycle methods to achieve the same functionality as a UID.

Creating a type field

Next, let's create a tutorial **type** field. We do not want this field to be free text as we are only going to support two types for a tutorial, so we are going to use the **Enumeration** field type. Proceed as follows:

1. Select the **Enumeration** field type.

2. Enter the name `type`.

3. In the **Values** text area, enter `video` and `text`. Make sure that each value is on a new line, as illustrated in the following screenshot:

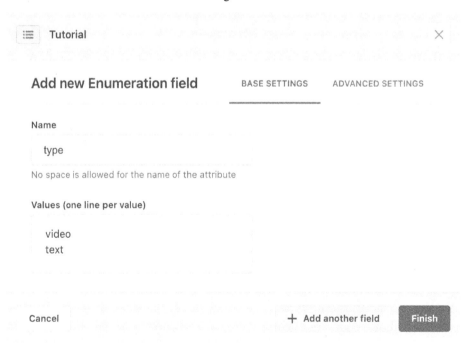

Figure 3.5: Enumeration field settings

4. Select the **ADVANCED SETTINGS** tab.

5. Choose video as the default value for this field.

6. Click **+ Add another field** to continue.

Creating url and contents fields

The remaining two fields are the **url** and **contents** fields. As the value for both fields is conditional depending on the tutorial type, we will just create the fields without any advanced settings. We will do the required validation when we do the customization later on in *Chapter 5, Customizing Our API*.

For now, let's choose **Text** for the **url** and **contents** fields. Proceed as follows:

1. Select the **Text** field type.

2. Enter the name as url.

3. Click **+ Add another field** to add the **contents** field.

4. Select the **Text** field type.

5. Enter the name contents.

6. Select the type **Long text**.

7. Click **Finish**.

> **What Is the Rich Text Field?**
>
> The **Rich Text** field is the Strapi equivalent of having a **What You See Is What You Get** (**WYSIWYG**) rich text editor. The editor will be available only via the admin panel, of course, and cannot be accessed through the API. Strapi uses Markdown to format the rich text.

At this stage, the **Tutorial** content-type is still not saved yet and no table is created in the database. Let's save the fields we have so far and we can come back to the field management screen later on to add the relation to the **Classroom** content-type.

8. Click the **Save** button.

9. Once you click the **Save** button, Strapi will translate those fields into a database table and database columns. The Strapi server needs to be restarted every time we make changes to the content-types and the database by definition.

You can see the process in action in the following screenshot:

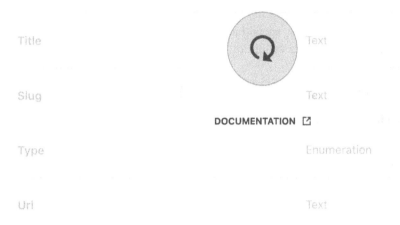

Waiting for restart...

You're using a feature that needs the server to restart. Please wait until the server is up.

Title   Text

Slug   Text

DOCUMENTATION

Type   Enumeration

Url   Text

Figure 3.6: Saving content-types and restarting the server

The last field remaining for our **Tutorial** content-type is the **Classroom** field. This field is going to be a relation to the **Classroom** content-type we created in *Chapter 2, Building Our First API*. However, before we create this field, we need to understand the different types of relations available in Strapi and how to use them. Let's do that in the next section.

Understanding relations in Strapi

When building an API, we often need to create relations between different entities of the system. Those relations range from simple and trivial relations to more complex ones. Strapi's **Relation** field allows us to establish a relation between multiple content-types.

Strapi provides us with six different types of relations to work with, as detailed next.

One-way

This relation reads as *has one*. If we choose this relation when creating a **Tutorial** content-type, the relation will read as *Tutorial has one Classroom*, and what will happen under the hood is that Strapi will create a **classroom foreign key** (**FK**) field in the **Tutorials** table in the database. When creating a **Tutorial** entity, we can associate it with one **Classroom** entity, and we will have a `classroom` object in the API response when we call one of the **Tutorial** GET API endpoints. The process is illustrated in the following screenshot:

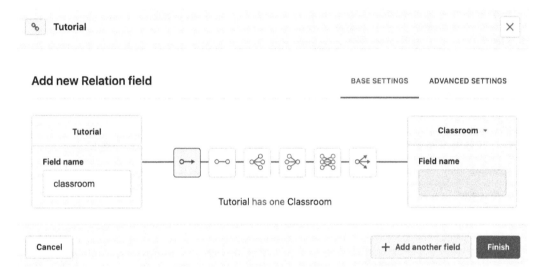

Figure 3.7: One-way relation example

One-to-one

This relation reads as *has and belongs to one*. If we choose this relation when creating a **Tutorial** content-type, the relation will read as *Tutorial has and belongs to one Classroom*. What will happen behind the scenes is that Strapi will create a **classroom** field in the **Tutorials** table in the database and a **tutorial** field in the **Classrooms** table. We can associate the two content-types while creating any of them. A `tutorial` object will be available in the **Classrooms** API response and vice versa for the **Tutorials** API response. The process is illustrated in the following screenshot:

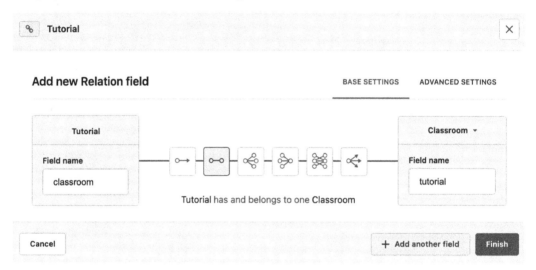

Figure 3.8: One-to-one relation example

One-to-many

This relation reads as *belongs to many*. If we choose this relation when creating a **Tutorial** content-type, the relation will read as *Tutorial belongs to many Classrooms*. What will happen behind the scenes is that Strapi will create a **tutorial** field in the **Classrooms** table in the database. When creating a tutorial, we can choose **multiple classrooms** to assign to; however, when creating a classroom, we can specify **only one tutorial** to be associated with. The **Tutorial** API response will include an array of classrooms while the **Classroom** API response will include a single `tutorial` object. The process is illustrated in the following screenshot:

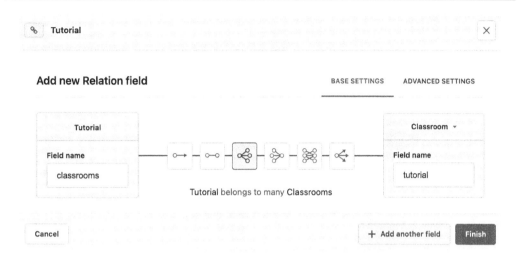

Figure 3.9: One-to-many relation example

Many-to-one

You can think of this relation as the opposite of the previous one-to-many relation we discussed. It reads as *has many*. If we choose this relation when creating a **Tutorial** content-type, the relation will read as *Classroom has many Tutorials*. What will happen behind the scenes is that Strapi will create a **classroom** field in the **Tutorials** table in the database. When creating a tutorial, we can choose one classroom to associate with, while we can choose multiple tutorials to associate with the classroom when we are creating a classroom. The **Tutorial** API response will include a `classroom` object, while the **Classroom** API response will include an array of tutorials. The process is illustrated in the following screenshot:

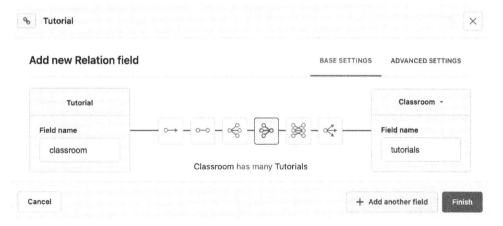

Figure 3.10: Many-to-one relation example

Many-to-many

This relation reads as *has and belongs to many*. If we choose this relation when creating a **Tutorial** content-type, the relation will read as *Tutorials has and belongs to many Classrooms*. Unlike other relations, Strapi will not alter any of the two tables involved in the relation; alternatively, it will create a **pivot table** to handle the many-to-many relation. When creating a tutorial, we can choose **multiple classrooms** to associate with it, and when creating a classroom, we can also associate **multiple tutorials** with it. The **Tutorial** API response will include an array of classrooms and the **Classroom** API response will include an array of `tutorials`. The process is illustrated in the following screenshot:

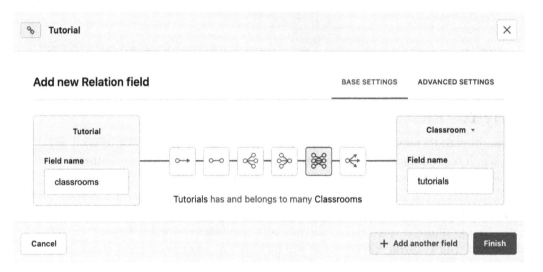

Figure 3.11: Many-to-many relation example

Many-way

The last relation reads as *has many*. If we choose this relation when creating a **Tutorial** content-type, the relation will read as *Tutorial has many Classrooms*. Similar to the many-to-many relation, a pivot table will be created for this relation. When creating a tutorial, we can choose **multiple classrooms** to associate it with. The **Tutorial** API response will include an array of classrooms, but no `tutorial` object will be shown in the **Classroom** API response since this relation is a uni-directional relationship. The process is illustrated in the following screenshot:

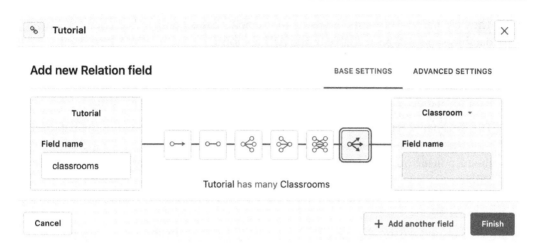

Figure 3.12: Many-way relation example

Now that we have learned the difference between relations in Strapi, let's use our new knowledge to choose the correct and appropriate relation to associate a tutorial with a classroom.

Creating a Tutorial and Classroom relation

We need to decide which of the six relations described in the previous section is suitable for associating the **Tutorial** and **Classroom** content-type. If you remember the system diagram in *Chapter 2, Building Our First API*, a teacher can create a tutorial and assign it to a classroom, and a classroom can have multiple tutorials. This means a classroom can have several tutorials associated with it. Based on this information, the many-to-one relation seems the most suitable one for this use case: **Classroom has many Tutorials**.

Going back to the **Tutorial** content-type, let's edit it and add the **Classroom** relation. Proceed as follows:

1. Click the **Add another field to this collection type** button.
2. Select the **Relation** field.
3. Select the fourth relation from the left. The text should read **Classroom has many Tutorials**. You should check the following:

   - The relation form shows the two content-types you are trying to associate.
   - The left-hand box shows the source content-type; this is the content-type we are editing.

- The field name represents the name of the field that is going to be created in the database and will be available in the API response for that content-type. In this case, it is **classroom**. We will leave the value as it is.

- The right-hand box shows the target content-type. We can use the caret down icon to change the content-type. We will leave the value as **Classroom**.

- The field name represents the name of the field that will be created in the database target table. If the relation requires a field to be created, it will also be included in the response body of our API when we query it for the **Tutorial** content-type. We will leave this value as it is.

4. Click **Finish**.

5. Click **Save** to save the changes. The server will restart again and we will have a relation.

Once we have the relation field created, we can go ahead and create test data to experiment with. Let's do that in the next section.

Creating a tutorial from the admin panel

Let's create a tutorial as test data. To do this, we need to navigate to the **Content Manager** using the left hand side menu and then select **Classroom** located under the **Collection Types** category. Proceed as follows:

1. Click **Add new tutorial**.

2. Fill the form with any sample data you want.

3. Notice the **Classroom** drop-down menu on the right-hand side. This drop-down menu allows us to assign the tutorial to a classroom. Click on it and choose one of the classrooms we created previously, as illustrated in the following screenshot:

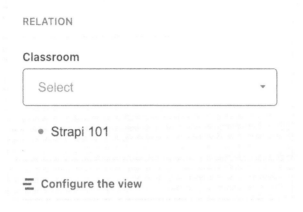

Figure 3.13: Assigning a relation in the Strapi admin panel

4. Click **Save** and then click **Publish**.

5. Feel free to add additional content if needed.

> **Note**
>
> We have created a slug manually here just for the sake of creating test data. Of course, this is not a practical approach, and, as discussed earlier in this chapter, we are going to change this in *Chapter 5, Customizing Our API*, when we start customizing the API and override the Strapi model and controller life cycle methods.

Now that we have data to play with, it's time to use Postman to see what the data looks like. But before we do that, we need to update the permission to allow us to access the `/api/tutorials` endpoints. Proceed as follows:

1. Click on **Settings** from the left-hand side menu.

2. Click **Roles** under the **User & Permissions** plugin.

3. Click **Public** to modify the permissions for the public role.

4.   Under **Permissions**, enable the permissions for **find**, **findOne,** and **create**, as illustrated in the following screenshot:

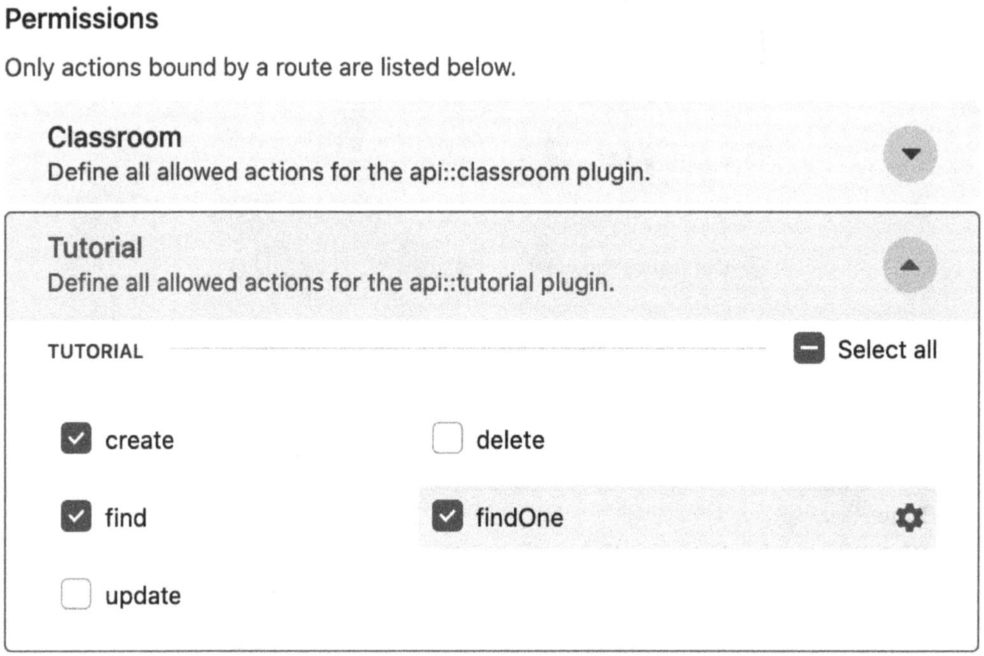

Permissions

Only actions bound by a route are listed below.

Classroom
Define all allowed actions for the api::classroom plugin.   ▼

Tutorial
Define all allowed actions for the api::tutorial plugin.   ▲

TUTORIAL ————————————————————————————————————   ⊟ Select all

☑ create   ☐ delete

☑ find   ☑ findOne   ⚙

☐ update

Figure 3.14: Updating Tutorial permissions

After we update the permissions, we can use Postman to interact with our newly added API endpoint.

Interacting with the API endpoints from Postman

Launch Postman if it was not running and try accessing the /api/tutorials endpoint. We should be able to see the tutorial data return in the API response; however, there is no classroom data returned in the response at all. This is because, by default, Strapi does not populate relations when fetching data, and we will need to explicitly specify that we would like the relation to be populated using the populate API parameter. So, let's update the URL by appending the populate parameter to it. The URL should look like this: /api/tutorials?populate=*. This time, we can see the tutorial data, including the **Classroom** relation.

Figure 3.15: /tutorials API endpoint response

What Is the `populate` Parameter?

The `populate` parameter is one of several parameters included with Strapi out of the box that we can use to customize the responses from the API. We will cover this in detail later on in *Chapter 6, Dealing with Content.*

Let's try another example by creating a new tutorial from Postman and sending a `POST` request to `/api/tutorials`. We will use the following payload:

```
{
  "data": {
    "title": "Content types in Strapi",
```

```
    "slug": "content-types-in-strapi",
    "type": "text",
    "contents": "Content types in Strapi allows us to
        define the entities that make up the api",
    "classroom": 1
  }
}
```

The preceding payload will create a new **text**-type tutorial and associate it with the classroom with the **identifier (ID)** equal to 1, as illustrated in the following screenshot:

Figure 3.16: Creating a tutorial using Postman

So far, we have been working with collection content-types, but what if we do not have multiple entries to work with? An example of this could be when we want to create static page content such as a home page for the frontend or storing API information metadata. We can use the single-type content-types in Strapi to achieve such a task, so let's see how to do it in the next section.

Differences between SINGLE TYPES and COMPONENTS

Strapi single types allow us to have one entry per single type. They are useful when we do not want to have multiple entries. We want to have an API name, a description, a version, and a list of developers as metadata information stored in a single type and have it publicly accessible via an endpoint. To do so, we will create a single type. Proceed as follows:

1. In the Strapi admin panel, navigate to the **Content-Type Builder** plugin page.
2. Click **Create new single type** under the **SINGLE TYPE** category.
3. We will call this collection **Info**.
4. Create **Short Text** fields for the **name** and **version** fields.
5. Create a **Long Text** field for the **description** field.

   The process is illustrated in the following screenshot:

Figure 3.17: API info Single type

Let's save the info Single type and add some content to it.

6. Click **Save** to save the Single type.

7. Wait for the server to restart. Once the page refreshes, you should see a new **SINGLE TYPE** category under the **Content Manager** menu item in the left-hand side menu.

8. Click on **Info**, as illustrated in the following screenshot, and populate the field with sample data:

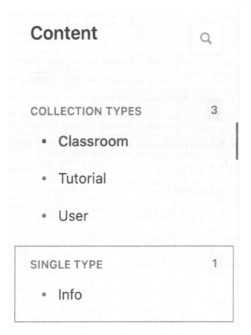

Figure 3.18: Editing API Info content

9. Click **Save** and then **Publish** to publish the changes.

10. Make sure to update the permission to enable a **find** action for the **Info** Single type.

11. Once you have updated the permission, you can access the new single type using the `/api/info` endpoint.

The final field we want to have in our **Info** content-type is a **developers** list. For each developer, we want to have a name and a GitHub repository URL. We can take advantage of Strapi components to achieve this.

Creating components

Components in Strapi are a combination of several fields; you can use components with both collection types and single types. Components are useful when you are managing the API contents from the Strapi admin panel as they allow you to have better-structured fields to edit the contents. Let's create a **developer** component and add it to our **Info** content-type. Proceed as follows:

1. In the **Content-Type Builder** plugin page, click **Create new component** under the **COMPONENT** section.

2. Enter **developer** for the name field.

3. The **Category** field is there to help you organize your components. Enter **information** for this field or any other values you want to use as a category name.

4. Choose any icon and then click the **Continue** button. You will be presented with a field selection form.

5. Create **name** and **github** fields, both using the **Text** type.

6. Click **Save** and wait for the server to restart.

7. Once the server restarts, you should see a new component called **developers** under the component list. Let's use it in the **Info** content-type.

8. Click the **Info** content-type to edit it.

9. Click **Add another field to this single type**.

10. Choose **Component** for the field type.

11. We will have a choice to create a new component if we do not have one or reuse an existing component. Choose **Use an existing component** to use our **developers** component.

12. Enter a name for the field we want to create in the **Info** content-type. We will use **developers**.

13. Select the **developers** component from the drop-down menu.

14. The **Type** field allows us to determine whether we want to have multiple instances or a single instance of the component. Since we want to have a list of developers, we will choose the **Repeatable component** option.

15. Click **Finish** to save the changes.

The process is illustrated in the following screenshot:

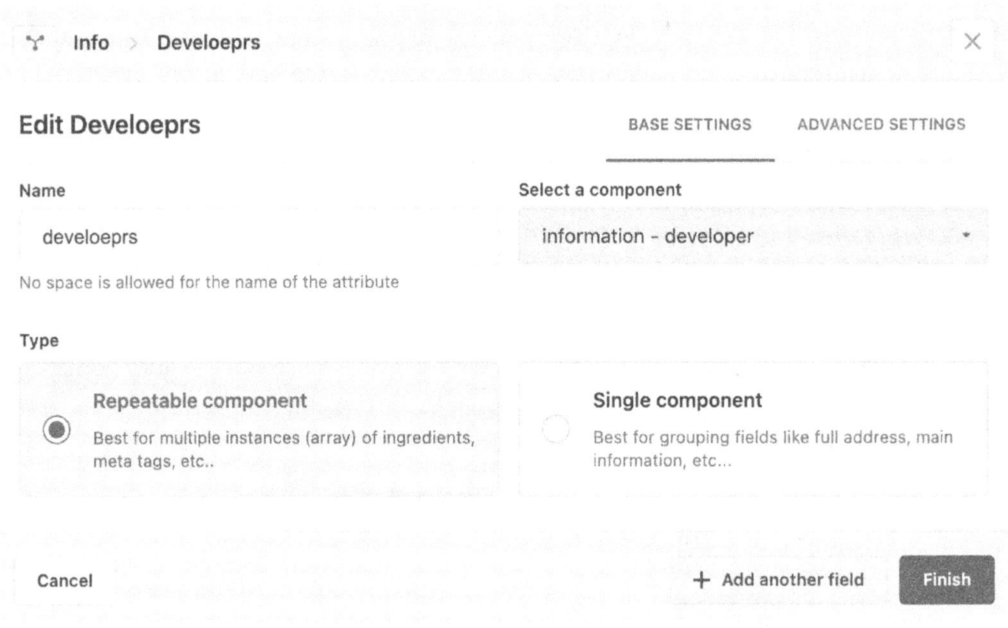

Figure 3.19: Adding the developers field to the Info content-type

16. Click **Save** to update the content-type and restart the server.

17. Let's edit the **Info** contents. We will switch to **Content Manager** by clicking on the **Content Manager** menu item and then clicking on the **Info** content-type from the left-hand side menu.

18. Notice the new **Developers** field.

19. Click **Add new entry** and enter sample data.

20. Save the changes.

21. Let's hit the `/api/info?populate=*` endpoint. This time, you should see a `developers` array field, as illustrated in the following screenshot:

Figure 3.20: The /info endpoint response

Summary

In this chapter, we explored the **Content-Type Builder** plugin in detail. We continued building on our API by adding a **Tutorial** content-type and deepened our knowledge of the **Content-Type Builder** plugin and the content-type fields in the process. We also saw how relations between content-types are managed in Strapi. We explored all six available relations in Strapi and created our first relation in the system between the **Classroom** and **Tutorial** content-types.

We then dug deeper into the different content-types in Strapi and illustrated by examples the difference between **collection types** and **single types**. Finally, we saw how to create custom components in Strapi and use these with content-types.

In the next chapter, we will have an overview of the Strapi admin panel. We will learn how to navigate our way through the admin panel, how to create new admin users and roles, and how to assign permissions to access the API to each role.

4

An Overview of the Strapi Admin Panel

In this chapter, we will explore the admin panel of Strapi. Strapi's powerful admin Panel is where we create content-types and manage content, permissions, plugins, and more. It is where we will spend a significant proportion of our time as developers, and where other stakeholders such as content creators and editors will interact with the product. The admin panel is very user-friendly and easy to explore and navigate, so we don't want this chapter to be a comprehensive manual for the admin panel; rather, it's a collection of tips and clarifications for some of the concepts of the admin panel so that you can use it more effectively.

These are the topics we will cover in this chapter:

- Navigating around the admin panel
- Managing admin users
- Managing content effectively
- Using the **Media Library** and consuming media in **API** (short for **application programming interface**)

Navigating around the admin panel

Once we log in to the admin panel, we can see a navigation menu on the left. The navigation menu is divided into several sections, as shown in the following screenshot:

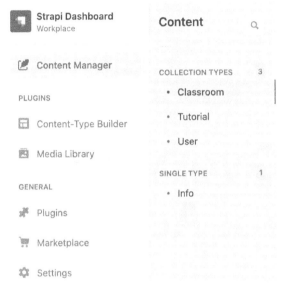

Figure 4.1: Strapi admin panel navigation menu

Let's look at the different sections of the navigation menu in more detail.

Content Manager

Content Manager is where we can manage the API content. **COLLECTION TYPES** are defined here. Out of the box, we have one type, **Users,** that is created by the **Roles & Permissions** plugin (covered in *Chapter 7, Authentication and Authorization in Strapi*). As we define more content-types in our project, they will appear here. Similar to collection types, **SINGLE TYPES** appear once we create our first single type in the Strapi **Content-Types Builder** plugin.

PLUGINS

By default, we have two plugins installed in Strapi: the **Content-Type Builder** plugin, which we used already in the previous two chapters to define our content-types, and the **Media Library** plugin, where we can add new media files—such as images, video, and document files—to use in our content.

GENERAL

Underneath the **PLUGINS** section, there is a **GENERAL** section containing some further settings, as detailed next.

Marketplace

In the **Marketplace** section, we are able to explore other community plugins and install them, as illustrated in the following screenshot. An example of such a plugin is the GraphQL plugin that we will install and use in *Chapter 10*, *Deploying Strapi*.

Plugins

Plugins lists the currently installed plugins and allows us to uninstall non-core ones. We will cover plugins in more detail in *Chapter 8*, *Using and Building Plugins*.

Settings

Settings, in our opinion, is a part of the admin panel that would benefit from some reorganization. At the moment, it seems like a catch-all section of the admin panel for seemingly unrelated functionalities. **Settings** is the place for the following actions:

- Managing admin users and roles
- Setting up internationalization options
- Setting up email providers
- Managing the **Users & Permissions** plugin

In the next section, we will explore the first of these functionalities: managing admin users and roles.

Managing admin users

One of the main use cases for the admin panel is to manage admin users. In Strapi, admin users and **Users** are different concepts. Let's explore the difference between the two before creating new admin users and assigning them roles.

The difference between Users and admin users

Users are a content-type created by the **Users & Permissions** plugin. These are the end users—the consumers—of our API. Normally, these will be created using the API itself and will have permissions to access certain actions on certain content-types. We will cover these in more detail in *Chapter 7, Authentication and Authorization in Strapi.*

Admin users represent the administrators of the whole Strapi instance. We already created our first admin user with a **Super Admin** role in the first step when we launched the admin panel for the first time and were prompted to create a first user. So, let's create a second admin user next.

Creating new admin panel users

To create a new admin user, follow these steps:

1.   First, head to **Settings** in the main navigation menu. Then, go to **ADMINISTRATION PANEL** and **Users**, as illustrated in the following screenshot:

Figure 4.2: Managing admin users under Settings | Administration Panel | Users

2. We then click on **Create new user**, which will open a form to create an admin user. Fill out the information you'd like for the user and click **Create user**, as illustrated in the following screenshot:

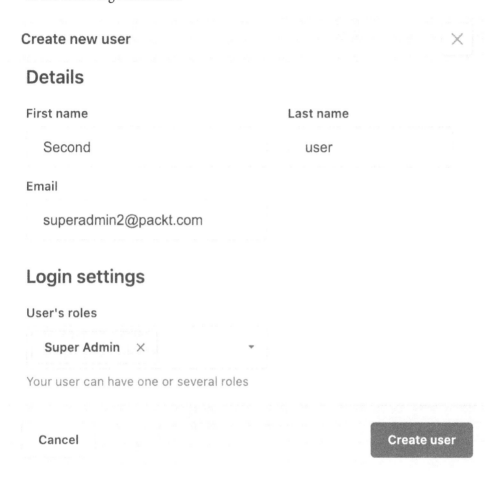

Figure 4.3: Form to create a new admin user

3. Once you save the user, you will be prompted with a link that you can copy and pass to your new admin user so that they can set up their account and choose their own password, as illustrated in the following screenshot:

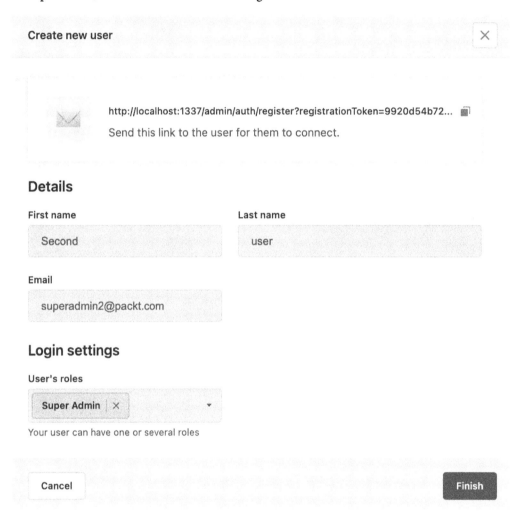

Figure 4.4: Generating a link to create a new admin user

For this new user, we chose the same role, **Super Admin**, as the first admin user we created when we set up Strapi. They can manage content-types, create new content-types, manage plugins, and basically perform every possible operation in the admin panel. Sometimes, we want to set up users with a different access level. This is where **roles** come into play.

Managing admin panel roles

Admin panel roles allow us to assign different levels of access for the admin users we created. By default, the admin panel comes with three different roles: **Super Admin**, **Author**, and **Editor**.

What these roles can do and can't do is configurable, as we will see in the next section. But Strapi, as usual, provides sensible defaults for us to use. By default, these are the three roles:

- An **Author** can manage the content they have created.

- An **Editor** can manage and publish content, including those of other users.

- A **Super Admin** can access and manage all features and settings.

The following screenshot shows where these roles are managed:

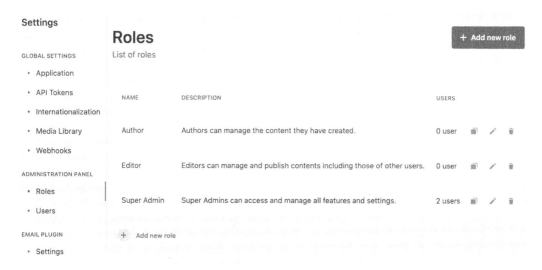

Figure 4.5: Managing roles under Settings | Administration Panel | Roles

Let's change the role of the admin user we created in the last step from **Super Admin** to **Editor** and then log in with that user.

Editing admin panel role permissions

Follow these steps to edit Admin Panel role permissions:

1. First, let's navigate to **Users** (under **Settings** | **ADMINISTRATION PANEL**) and click on the admin user we created in the previous section. Then, change the role of the user to **Editor**, as illustrated in the following screenshot:

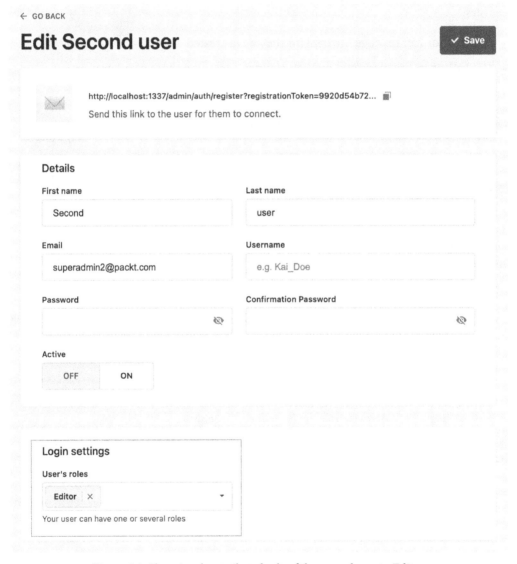

Figure 4.6: Changing the attributed role of the second user to Editor

2. Then, open a new browser (or incognito/private window) and log in as the user we just edited. You will observe that the admin panel looks rather empty. No content-types are showing on the navigation menu, nor any other sections, as the following screenshot illustrates:

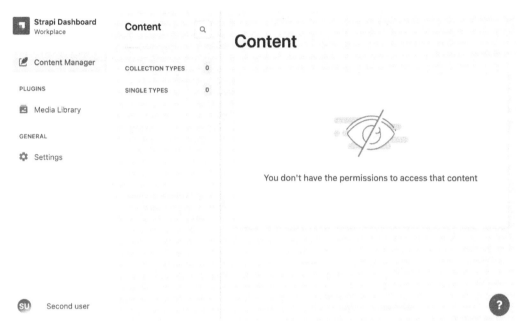

Figure 4.7: No content-types showing when logging in as the Editor user

The reason is that **Editor** users are not yet allowed to view any content-types. Similar to API users, contents are inaccessible by default.

3. To be able to give them permissions, we can log back in as the **Super Admin** user, navigate to the **Roles** section (under **Settings | Administration Panel**), and then grant that role (not the user) permission to manage a certain content-type.

4. We can edit the **Editor** role to give it permission to manage the **Classroom** and **Tutorial** content-types, as illustrated in the following screenshot:

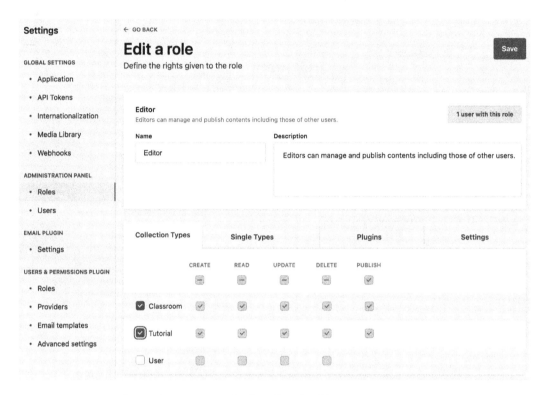

Figure 4.8: Updating the permission for the role to be able to manage certain content-types

5. Once we log back in as the **Editor** user, we will notice that we can now see the **Classroom** and **Tutorial** content-types on the left-hand side (but not the **Users** content-type), as illustrated in the following screenshot:

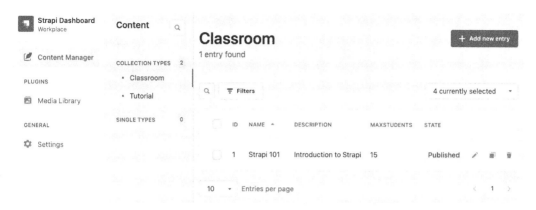

Figure 4.9: After updating the permission, Classroom and Tutorial appear for the Editor user on the left-hand side

---

Limitations of the Community Edition

Sadly, in the Community Edition, we're not able to add more admin roles. We also have a limit in terms of the *granularity* with which we can configure the predefined roles; for example, we can grant **Author** and **Editor** roles access to all the operations related to a content-type, similar to what we did in the previous section when we allowed **Editor** roles to perform all **CRUD** (short for **create, read, update, delete**) and publish operations on the two content-types. We were not able to define a more granular set of permissions—for example, to grant them permission to create but not delete. This level of granularity is not available in the Community Edition.

Having understood how to add new admin panel users, as well as the difference between API users and admin panel users, we will now explore the **Content-Type Builder** plugin and learn a few tips to use it effectively.

Managing content effectively

We already spent a significant amount of time in the **Content Manager**, adding **Classrooms**, **Tutorials**, and other entities to our system. While many applications might have custom interfaces to edit these content-types (imagine a React application where **Users**—not admin users—can sign in and manage entities), in many scenarios, you will only manage content through the admin panel. Other than us, the **Super Admin** users of the Strapi instance, we learned in the previous section that we can also have **Editor** and **Author** users who can manage the content as well. In a typical organization, this can be the marketing department, content editors, and other stakeholders.

Once our content starts growing and we have hundreds or even thousands of entities, it becomes tedious to just browse through a long list of table rows. Strapi has some handy functionalities to explore, find, and manage content easily.

Searching and filtering to query our data

Strapi allows us to search for content easily through the *search* functionality. There are two options for searching for entities. The first is the **Search** textbox, as illustrated in the following screenshot:

Figure 4.10: At the top of the admin panel, you can use the search box to perform an ad hoc search for all the entities of the current content-type

This search performs a "fuzzy" search against all the fields of the content-type. As our database grows and the number of entries increases, this might become slower. In those cases, or when we want to perform a more specific search, we can then use filters.

To display filters, we can click on the **Filters** icon on top of the table view of the current content-type; then, we can create any query we like. For example, we can create a filter to show only the entries with maxStudents greater than 3, as illustrated in the following screenshot:

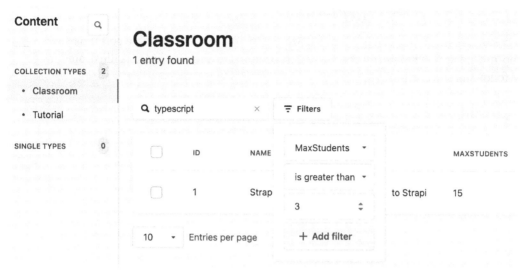

Figure 4.11: Using filters to query your data more specifically and combining filters

We can also combine multiple filters with an AND operator by clicking on the + icon and applying additional filters.

> **Under the Hood: What Is SQL?**
>
> Strapi uses the Knex library under the hood. Knex is a database-agnostic **Structured Query Language (SQL)** builder, so it's not surprising that all the API calls and the actions that we perform on the admin panel are transformed eventually into a SQL query (in the case of relational databases). If you are interested in knowing (or debugging) which SQL queries are being generated by Strapi and Knex, then you can run Strapi with an extra flag: DEBUG=knex:query yarn develop. This will run Strapi in debug mode and set it to print queries from Knex. We will see in the logs all the queries that run against the database when we hit the API or when we navigate through the admin panel (see *Figure 4.12*).

Figure 4.12: Running in debug mode allows us to check which SQL queries are being run by Strapi

Customizing the table view

To customize the table view in **Content Manager**, you can use the table columns drop-down list to *temporarily* choose which columns to show or hide, as illustrated in the following screenshot:

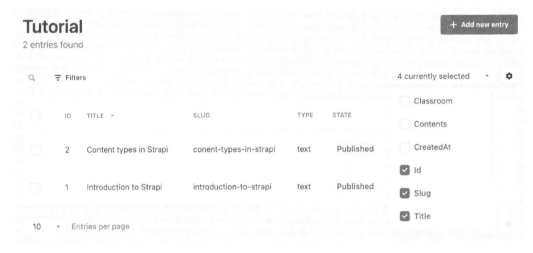

Figure 4.13: Using the Configure button on the top right of the table to show or hide columns in the table

The changes applied to the table will be lost when we refresh the page. If we want the configuration to be persisted, we can click on the **Configure** icon button, followed by **Configure the view**. This will take us to a page where we can configure the table view columns, their order, the sort of attribute for the content-type, and much more, as illustrated in the following screenshot:

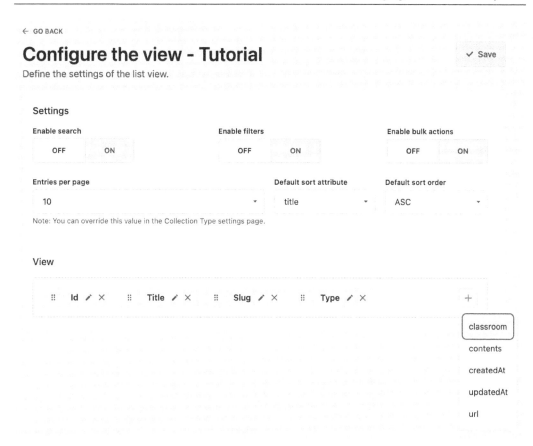

Figure 4.14: Using Configure the view to apply changes to the table that will be persisted and applied every time you open this content-type in this environment

Changes we make on the **Configure the view** page will be persisted and applied every time we view this content-type in the admin panel.

> **Note**
> Changes applied to the view are persisted in the database so that they are applied only in the current environment we're working on. We will talk about migrating data and settings between environments and some of the considerations we should keep in mind in *Chapter 9, Production-Ready Applications*.

Customizing the details view

The details view (when you click on a single entry in the table view) can also be customized with different options to make it look and be organized the way you prefer, as illustrated in the following screenshot:

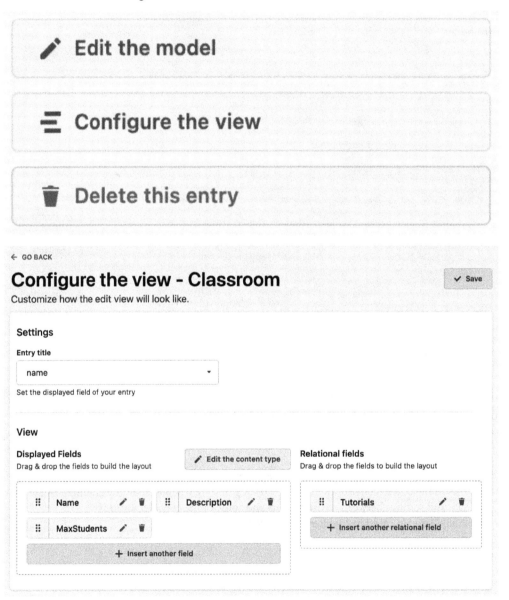

Figure 4.15: Clicking on Configure the view will take you to a screen to configure how the details view of an entity should look

Here are some ways in which you can configure the view:

- Choose the field to be used as the entry title. This is handy when we want to set a relationship (between **Tutorial** and **Classroom**, for example) to set which field will be displayed when setting up the relationship. (Note that internally in the database, the **identifier** (**ID**) of the entity will be used to set up the relationship, but this is about which field to show in the admin panel.) You can see an example of this in the following screenshot:

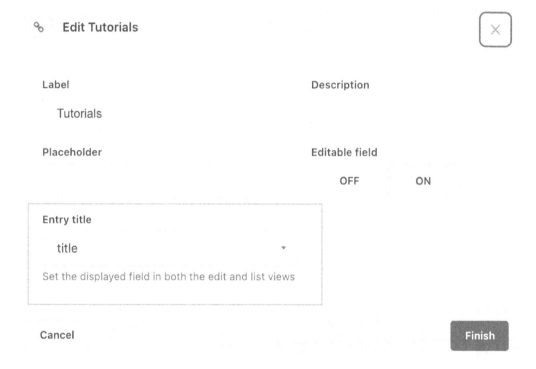

Figure 4.16: Example of configuring an entry title

- Choose which fields to display and their order:

  a. You can hide some fields from the admin panel—for example, if you want them set up in the background based on some other criteria (that is, a **Summary** field generated automatically out of the 50 first words in the **Description** field).

  b. You can reorder the fields in the admin panel.

- Customize a single field, including changing its label, setting a description and a placeholder, or disabling editing, as illustrated in the following screenshot:

| Ab   **Edit Name**   ✕ |

Label   Description

  Name

Placeholder   Editable field

                                     OFF   ON

Cancel   Finish

Figure 4.17: Clicking on a single field, you can customize the label, description, and placeholder for that field

The admin panel is very customizable and user-friendly. It is constantly improving with every release of Strapi. The topics and tips we covered here will help you make better use of it, especially once the amount of content you have starts to grow. Next, we will explore another part of the admin panel: the **Media Library**. In most systems nowadays, some of our content will not be text-based, so it is important that Strapi supports a way to manage media content such as images, videos, and documents.

Using the Media Library

The **Media Library** allows us to manage media content such as video, images, and documents inside the admin panel. This can be done easily by navigating to the **Media Library**, clicking on **Upload assets**, and then either selecting files from your PC or providing **Uniform Resource Locators** (**URLs**) of the media files you want to upload. You can see a screenshot of the **Media Library** here:

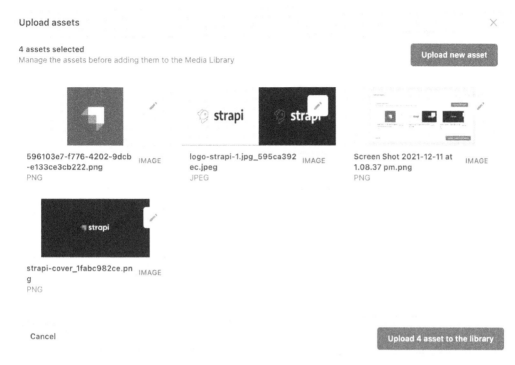

Figure 4.18: In the Media Library, you can select which assets you want to upload to Strapi

Once we upload the media files, we can then make use of these in our other content-types. So, let's head back to our **Tutorials** content-type and add a new field, `coverImage`, to this content-type.

We head to **Content-Type Builder**, choose **Tutorials**, and then **Add another field to this collection type**. Select the field type as **Media**, and set the options for our new field, as follows:

- **Name**: `coverImage`
- **Type**: **Single media**

The process is illustrated in the following screenshot:

Figure 4.19: Adding a new Media field to our Classroom content-type

There are two types of **Media** fields: **Single media** and **Multiple media**. As the name and the description explain, multiple media is an array of media files suitable for cases such as multiple images for a slider.

- Under **ADVANCED SETTINGS**, we can also set more options. One useful setting in our case is to restrict the allowed media files to images only, as we will not be accepting videos or files in our case.

Under **Select allowed types of media**, uncheck **All** and check **Images**, as illustrated in the following screenshot:

Figure 4.20: Under ADVANCED SETTINGS, we can change more settings for the fields—for example, restrict which media types are allowed

Once we edit the options, click on **Finish**, then **Save** again (to save the content-type). The Strapi instance will be restarted since we changed a content-type. Now, if we navigate to the **Tutorials** content-type and either edit or create a new entry, we will see a new field, **CoverImage**, where we can select an image from our **Media Library** as the cover image for our entry. The field is shown in the following screenshot:

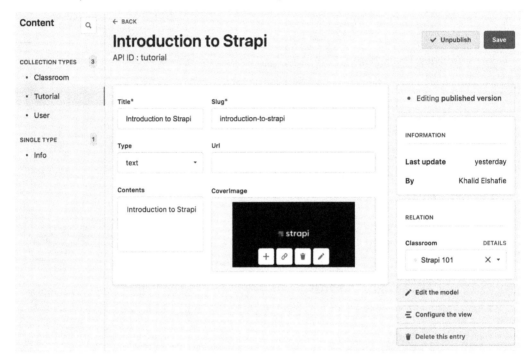

Figure 4.21: Selecting a cover image from the Media Library for our Tutorial entry

Now that we have set up the field in the admin panel and added it to some of our content, let's see what it looks like in the API response.

Media types in API responses

Let's go back to Postman and hit the API to get information for the **Tutorial** entry we just edited. We can call GET /api/tutorials?populate=* to get all tutorials, or GET /api/tutorials/tutorialID?populate=* (replace tutorialID with the actual ID of the entry in Strapi).

> **Got 403?**
>
> If you receive a `403 Forbidden message`, remember to update the permissions for the **Tutorial** content-type to allow unauthenticated requests (under **Settings** | **Users & Permissions** plugin | **Roles**, check the **Tutorial** `find` and `findAll` actions, then click **Save**). This was covered in detail in *Chapter 2, Building Our First API*.

The response from the API will contain our new `coverImage` field, as illustrated in the following screenshot, but it's not just a simple reference to an image URL:

Figure 4.22: The response to GET /api/tutorials/1 contains our new coverImage Media field

Strapi optimized the image we uploaded, and the API returns several optimized versions in the response.

How Strapi optimizes media files automatically for us

Strapi automatically created *four* versions of our image with different sizes: *thumbnail*, *medium*, *small*, and *large*, as well as the *original* image. This is incredibly helpful as large images and media files are often the culprit in having large loading time for web applications, and lead to sluggish experiences on mobile applications. By having optimized versions of the media file created for us automatically, clients of the API can choose which version to use and optimize the **user experience** (**UX**); why load the original large high-definition image when we're only showing a small avatar in a list, for example? We can simply use the thumbnail version Strapi generated for us in this use case.

What If I Want to Turn Off This Behavior?

While we can't see a reason why you would want to turn off this optimization behavior, we can switch it off from **Settings** | **Global Settings** | **Media Library**. You can also enable **auto orientation** in the settings, which is turned off by default—this will automatically orient your images in the right direction when uploaded.

By default, the uploaded files themselves are saved in the filesystem, under our Strapi project, in the /public/uploads folder, as illustrated in the following screenshot. If you are using an ephemeral filesystem, as in *Heroku*, then these uploaded files will be wiped out on every restart of the server. We will discuss ways to customize this behavior (to save to Amazon **Simple Storage Service** (**S3**), for example) in *Chapter 9, Production-Ready Applications*.

Figure 4.23: The uploaded media is saved under the public/uploads folder

Summary

In this chapter, we stepped out of code-land a bit to spend some time around the admin panel. In a typical organization, a **content management system** (**CMS**) has many stakeholders: developers are typically responsible for designing the system's entities (that is, deciding on which content-types make up part of the system; the relationships between entities; the permission model; any custom code; and so on). Other stakeholders—such as marketing, content editors, and creators—are responsible for creating the content itself. Regardless of our roles, when dealing with Strapi, the admin panel remains the place where we will spend a considerable amount of time, so it's important to be comfortable and efficient when using it.

To achieve this efficiency, we revisited **Content Manager** and saw how the views can be customized to users' needs. We also used **Media Manager** and saw how a content-type can make use of media files, and how they are represented in the API.

Other than managing content, we created new admin users and assigned them different roles. More importantly, we also explored the difference between admin users and API users. This is often quite confusing when starting with Strapi.

This chapter is the end of the first part of the book, which has focused on understanding the core concepts of Strapi. Next, we will start customizing our APIs beyond the CRUD operations that we got out of the box so far using the admin panel.

Section 2: Diving Deeper into Strapi

In Section 2 of the book, we will have a closer look at how Strapi works. We will learn how to customize the API, understand the life cycle methods of Strapi components and how authentication works, and finally, see how to work with plugins in Strapi, as well as creating our own plugin.

In this section, we will cover the following topics:

- *Chapter 5, Customizing Our API*
- *Chapter 6, Dealing with Content*
- *Chapter 7, Authentication and Authorization in Strapi*
- *Chapter 8, Using and Building Plugins*

5
Customizing Our API

In this chapter, we will start customizing our API beyond basic **create, read, update, and delete** (**CRUD**) operations. So far, all we have done is define content-types in Strapi's Content-Type Builder, which automatically created an API to create, retrieve, update, and delete entries for that type. We did not have to do any coding. As always, Strapi provided sensible defaults and got our project off the ground with minimum effort. In any large project, though, we will reach a point where these defaults are not enough, and we will want to do a bit more or alter the default behavior slightly. Strapi is very easy to extend, and this chapter will teach us how to do so.

In this chapter, we will cover the following topics:

- Defining new routes in our API
- Mapping routes to controllers and services
- Implementing the DRY concept for our API logic
- Securing our API by removing sensitive data and configuring the default permission policy
- Overriding the default behavior of Strapi CRUD APIs
- Using life cycle hooks to perform operations before and after CRUD events

The routes – where it all starts

In *Chapter 2, Building Our First API*, we briefly described the components that make up a Strapi API. The first of these, from an API consumer point of view, is the **route**. Routes define the external interface to our API; that is, the URIs the consumers of the API need to communicate with to interact with the API:

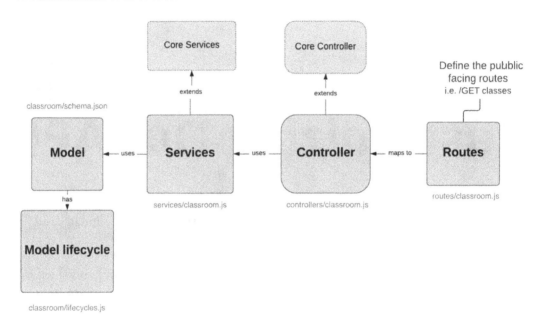

Figure 5.1: Components of a Strapi API

By default, when we create a content-type, Strapi creates a REST API with core routes to perform CRUD operations. The definition for these routes exists in code and can be changed in code instead of using the admin panel or the Strapi CLI, as we have done so far.

Let's look at the `tutorial` type we created in the previous chapters and the code that was generated in more detail to understand what Strapi gave us by default.

The default routes

When we defined the `tutorial` content-type in the admin panel, Strapi built a RESTful API for our content-type. It defined the `createCoreRouter` factory function, which automatically generates five core routes to perform CRUD operations (that is, `find`, `findOne`, `create`, `update`, and `delete`). `createCoreRouter` is defined in `{content-type}/routes/{content-type}.j` (`tutorial/routes/tutorial.js`). If you take a look at the file, you will see that it's empty except for the definition of the `createCoreRouter` factory function. As you may recall from *Chapter 2, Building Our First API*, we explained that most of the generated files are empty because they inherit their behavior from the core components provided by Strapi. These files are there in case we want to override one of these components, which we are going to do later in this chapter.

In the case of our `tutorial` content-type, these are the routes that will be generated by default:

HTTP Verb	Path (URI)	Default Handler (Defined in Controllers)	Operation
GET	/api/ tutorials	`tutorial.find`	Gets all tutorials
GET	/api/ tutorials/:id	`tutorial.findOne`	Gets a single tutorial by ID
POST	/api/ tutorials	`tutorial.create`	Creates a new tutorial
PUT	/api/ tutorials/:id	`tutorial.update`	Updates the tutorial that matches the specified ID
DELETE	/api/ tutorials/:id	`tutorial.delete`	Deletes the tutorial that matches the specified ID

> **Listing All Routes**
>
> If you would like to list all the application routes, you can use the Strapi CLI `routes:list` command to do so. The full command is `yarn strapi routes:list`.

As always, Strapi used sensible defaults for us. The default API conforms to the REST architectural style, but since all of these URIs are defined in code, we can add new ones, change the verbs, or delete them altogether.

> **Why are Routes Defined in Code?**
>
> Strapi prides itself on being developer-friendly. The fact that these routes are defined in code, as opposed to being defined in a database table, for example, gives us a lot of flexibility, as we shall see shortly. It also means that the history of our API lives in code, under source control, and has the same quality controls as any other source code. This decision also makes it easy to move the API between environments; just push the code and all our routes and the models behind them are defined in the new environment.

To illustrate how routes work, let's imagine that we need to get rid of one of these routes:

As an API user

I want to disable listing all tutorials

So that they are not accidentally exposed

This is a common security requirement in many projects. While on a REST API, consumers might be expecting that GET /api/tutorials returns all tutorials, so it's not uncommon to sometimes do things that are not 100% RESTful, especially when we are building an API for a specific consumer or use case.

Disabling a core route

Before we attempt to fulfill this requirement, let's try to understand the createCoreRoutes factory function first. The createCoreRoutes factory function accepts two parameters:

- The first one is a uid string for identifying the routes, whose format is api::api-name.route-name.

- The second parameter is a configuration object that allows us to override the core route configurations.

The following code snippet shows the configuration object properties:

```
module.exports = createCoreRouter('api::tutorial.tutorial', {
  prefix: '',
  only: [],
  except: [],
  config: {},
});
```

This configuration object contains the following properties:

- `prefix`: This allows us to add a custom prefix to all the model routes. For example, if we set the value to `/test`, then all of the tutorial's endpoints will have `/test` in the URL. For example, the get-all tutorials endpoint will be `/api/test/tutorials`.

- `only`: This is an array that allows us to specify which routes we want to be available. In other words, instead of having all five core routes available, we will only have the routes specified in this array.

- `except`: The `except` parameter is the opposite of the `only` parameter. If we use this configuration option, then all five core routes will be available, except for the routes defined in this array.

- `config`: The `config` route allows us to configure route middleware, policies, and public availability. We will discuss this later in *Chapter 7, Authentication and Authorization in Strapi.*

Going back to our requirement, we can use the `except` configuration to disable listing all the tutorials. This change is shown in the following code snippet:

```
module.exports = createCoreRouter('api::tutorial.tutorial', {
    except: ['find'],
});
```

Once we save this change, if we attempt to access the `GET` `/api/tutorials` endpoint, we will get a `404` response code since the `find` endpoint is no longer available:

Figure 5.2: A 404 response code after disabling the endpoint

Disabling a core route is more likely to be a requirement. For example, for security reasons, we might decide that we don't want an endpoint that lists all tutorials, or we only want to access tutorials as part of the `classrooms` content-type. In that case, we can disable the endpoint altogether as we have no use for it and we don't want to accidentally expose it.

Now that we have disabled a route, let's look at adding a new route to our API and see how that new route links to other components of a Strapi API, namely `controllers` and `services`.

Adding a new route

We have a new requirement for our API:

As an API user

I want to get all the tutorials associated with a classroom

So that I can know what tutorial to follow

We just deleted the `GET /tutorials` endpoint, and while calling `GET /classroom/:id?populate=*` also returns the list of tutorials associated with the class, the primary entity that's returned is a `Classroom`. We want to add a new route to be able to get the tutorials associated with a `classroom`.

We will call the new route, `GET /classrooms/:id/tutorials`, which will return a list of tutorials (not classrooms with tutorials) associated with a `classroom` with that ID.

To add the new route, we will need to create a custom route. Custom routes can be defined by creating a file in the `{content-type}/routes/` folder. The filename itself does not matter. However, as a best practice, it is always recommended to use meaningful and descriptive names. We will use `custom-classroom.js` as the file name. The full path of the file is `src/api/classroom/routes/custom-classroom.js` and its content is as follows:

```
module.exports = {
  routes: [
    {
      method: 'GET',
      path: '/classrooms/:id/tutorials',
      handler: 'classroom.findTutorials',
      config: {
        auth: false,
      },
    },
  ],
};
```

Figure 5.3: The classroom custom route file

When creating a custom route, we will need to export a named `routes` array with the following properties:

- `method`: One of the GET, POST, PUT, or DELETE HTTP methods.

- `path`: The URI of the route.

- `handler`: The controller's method that will handle the request. It takes the form of `controller.method`. This points to a method that's been defined in the controller module in `{content-type}/controller/{content-type}.js`.

- `config`: This is where we can pass an array of policies, middleware, and public availability. In our case, we set this route to be publicly available by specifying the `auth:false` value. We will cover this in detail later in *Chapter 7, Authentication and Authorization in Strapi*.

Once we save the file, Strapi will reload automatically to pick up changes since we are in development mode. Then, it will crash with an error stating `Error creating endpoint get /classrooms/:id/tutorials: handler not found "classroom.findTutorials"`:

```
[2021-12-12 12:44:48.295] debug: ● Server wasn't able to start properly.
[2021-12-12 12:44:48.296] error: Error creating endpoint GET /classroom/:id/tutorials: Handler not found "classroom.findTutorials"
Error: Error creating endpoint GET /classroom/:id/tutorials: Handler not found "classroom.findTutorials"
    at getAction (/Users/khalid/code/nyala/Building-APIs-with-Strapi/node_modules/@strapi/strapi/lib/services/server/compose-endpoint.js:131:11)
```

Figure 5.4: Strapi throws an error as it can't find the route handler

To fix this error, we will need to define this handler in our controller. So, it's time to leave the route definition files and dig a level deeper into controllers and services.

Handling routes with controllers

The **custom routes** definitions (in `src/api/tutorial/routes/custom-tutorial.js`) have two main responsibilities:

- To define how the external world can call an API operation
- To map that operation to an internal method

The method we map to lives in a component called `Controller`. This mapping is set in the `handler` property in the route definition (they are referred to as *actions* in the Strapi documentation).

`controller` encapsulates a set of methods that we can point to from our API endpoints. It's almost a one-to-one relationship since each route has a handler function (although in theory, nothing prevents us from pointing many routes to the same handler).

Each content-type has its controller set up in the module in `src/api/{content-type}/controllers/{content-type}.js`. For the `classroom` content-type, the controller is defined in `src/api/classroom/controllers/classroom.js`.

Let's open the file and explore its contents.

Adding a new controller handler for our endpoint

Similar to what we saw with the tutorial core route file, the classroom controller file is almost empty right now, except for the `createCoreController` factory function. This factory function, as you may have guessed, generates the `core` controller actions and allows us to override those actions, as well as define a custom action.

The `createCoreController` factory function accepts two parameters. The first one is a `uid` string for identifying the controller and its format is `api::api-name.controller-name`. The second parameter is a callback that returns a configuration object that allows us to override the core controller actions or add new ones.

Our API crashed as we assigned a `classroom.findTutorials` handler to our new route. Since this is not a handler that's defined in the `Core` controller, we need to add this controller action ourselves.

So, let's extend our `Classroom` controller by adding this new handler:

```
"use strict";

/**
 * classroom controller
```

```
*/
const { createCoreController } = require('@strapi/strapi').
factories;
module.exports = createCoreController ('api::classroom.
classroom', ({ strapi}) => ({
  async findTutorials() {
    return "to be implemented";
  },
});
);
```

Please note that the route definition we added in the previous section had its `handler` value set to `classroom.findTutorials`. This follows the `controller.method` format, so the first part classroom matches the controller (the filename), while the second part, `findTutorials`, matches the method we just added.

Now, if we restart our crashed server (`yarn develop`), it will start successfully. If we head to Postman and hit the GET `/api/classrooms/1/tutorials` endpoint, we will get a success *200* response, along with our hardcoded string:

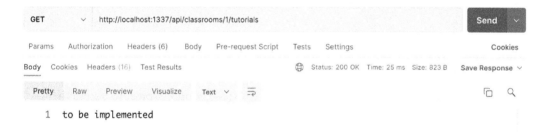

Figure 5.5: Our new endpoint now returns our hardcoded string

A *200* response is not appropriate in this case since the result is not a success. Let's see how we can change it to something more appropriate.

Controllers – it's all about context

All the handlers in our controller receive a context object as a first argument. As we mentioned in *Chapter 1*, *An Introduction to Strapi*, Strapi is built on top of the popular library **KOA**, so we can manipulate this context object in all the ways that **KOA** allows us to manipulate the response or the request object. For example, if we want to return a 501 HTTP status code for now, then we can simply change the context response object, like so:

```
async findTutorials(ctx) {
    ctx.response.status = 501;
    return "to be implemented";
}
```

Now, hitting the endpoint will return 501 (not implemented) instead:

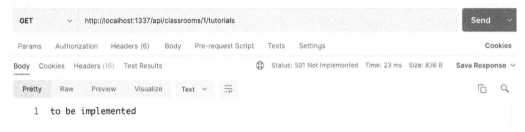

Figure 5.6: We manipulated the response context to return a custom HTTP status code

Accessing params and query objects from the context

Other than manipulating the response, we typically use the context object to access the query and params properties of the request. **Params** are defined in the route definition (prefixed by a colon; that is, `:id`). Query params are more flexible and appear after a question mark in the route (that is, `/tutorials?title=Strapi`). **Query** properties are more flexible as they don't need to be defined beforehand in the route definition and we can have as many of them as we like, separated with an `&`.

Let's amend our handler to log the values for `params` and `query` to the console:

```
async findTutorials(ctx) {
    ctx.response.status = 501;
    strapi.log.debug(`params:
    ${JSON.stringify(ctx.params)}`);
    strapi.log.debug(` query: ${JSON.stringify(ctx.
    query)}`);
```

```
    return "to be implemented";
}
```

Now, if we hit the `GET /api/classrooms/1/`
`tutorials?publisher=packt&name=Strapi` endpoint, it will still call our
handler, and the logs that have been printed in the console will show these values:

```
debug: params: {"0":"api/classroom/1/tutorials","id":"1"}
debug: query: {"publisher":"packt","name":"Strapi"}
http: GET /api/classroom/1/tutorials?publisher=packt&name=Strapi (13 ms) 501
```

Figure 5.7: The API now returns 501

> **Using strapi.log**
>
> Instead of using `console.log`, we used `strapi.log` to log the values
> for `params` and `query`. `strapi.log` contains a reference to Strapi's
> logger, which uses the *Pino* library, a low overhead Node.js logger that can be
> used in production while also providing benefits such as pretty formatting and
> other capabilities in development.

Now that we have a new route pointing to our custom handler, we need to implement its
functionality: we must get the list of tutorials related to that classroom. To do so, we will
need to make use of **services** that allow us to query the database and return what we want.

Reusing logic with Strapi services

Services are utility functions that are mostly used by controllers to perform different
operations. Their typical use is to access the database, but they can encapsulate any other
logic – such as sending emails or integrating with other systems – that we want to share
with other components of a Strapi API. Conceptually, they keep our logic **DRY** (which
stands for **don't repeat yourself**) by encapsulating these reusable operations.

Similar to controllers, if we have a look at the service file that Strapi generated for us (for
example, `services/classroom.js`), we will find that it's empty. This is because,
similar to the routes and the controller, Strapi provides a default `Core` service that we
can customize or extend in this file if we want to (the same concept as the default `Core`
controller in the previous section) using the `createCoreService` factory method.

Let's customize the classroom service by adding a new method called `findTutorials`. For now, we will just use a simple debug statement. The full code is shown here:

```
'use strict';
/**
 * classroom service.
 */
const { createCoreService } = require('@strapi/strapi').
factories;
module.exports = createCoreService(
'api::classroom.classroom',
({ strapi }) => ({
  findTutorials(classroomId) {
    strapi.log.debug(`findTutorials: classroomId =
    ${classroomId}`);
  },
})
);
```

Next, let's update the classroom controller so that it uses the newly created `findTutorials`. To use a method from a service in a controller, we must access it through `strapi.sevice`. The updated controller code is illustrated in the following code snippet:

```
async findTutorials(ctx) {
    const {params} = ctx;
    await strapi.service('
    api::classroom.classroom').findTutorials(params.id);
    return "to be implemented";
}
```

If we try to access the GET `/api/classroom/1/tutorials` endpoint now, we should be able to see the debug log from our classroom service in the console:

```
debug: findTutorials: classroomId = 1
http: GET /api/classroom/1/tutorials (12 ms) 200
```

Figure 5.8: Log message from the classroom service

So far, we can invoke our `findTutorials` service method from our classroom controller and print a simple debug statement to the console. However, we are still unsure of how can we perform a query against our database to get tutorials associated with the classroom ID that's passed to the `findTutorials` service method. By looking at the Strapi documentation, we can see examples of how to override some of the core default implementations. We can see that Strapi exposes the **Entity Service API**, which allows us to communicate with the database. Now, let's learn how to use the **Entity Service API** to customize our tutorial service.

Communicating with the database using the Entity Service API

The **Entity Service API** is the layer in Strapi that's responsible for managing Strapi's complex data structures, such as **components**, which were introduced in *Chapter 3, Strapi Content-Types*, as well as performing database queries using the **Query Engine API** under the hood.

> **What is the Query Engine API?**
>
> The Query Engine API is a layer in Strapi that allows you to interact with the database at a lower level. It gives unrestricted internal access to the database layer. We will discuss the Query Engine API in more detail in *Chapter 6, Dealing with Contents*.

The **Entity Service API** can be accessed through `strapi.entityService`. It exposes multiple methods (`strapi.entityService.[findOne, findMany, create, update, delete]`) to preform **CRUD** operations on entities.

For our requirements, we want to get all the tutorials associated with a specific classroom. Using the `findMany` method will allow us to satisfy this requirement as it will allow us to retrieve multiple entities using certain conditions. Let's update the `findTutorials` method in the classroom service, as follows:

```
module.exports = createCoreService(
'api::classroom.classroom',
({ strapi }) => ({
  findTutorials(classroomId) {
  return strapi.entityService.findMany('
  api::tutorial.tutorial', {
  filters: { classroom: classroomId },
```

```
    });
    },
  })
);
```

The `findTutorials` method we just created expects `classroomId` as a param. It will use this param to query the data using the **Entity Service API**. The `entityService.findMany` method accepts two parameters – the first is a `uid` string in the `api::api-name.conent-type-name` format, while the second is a configuration parameter to manipulate the result. In the previous code snippet, we only used one parameter with the `findMany` method, namely the `filters` parameter. The `filters` parameter, as its name suggests, allows us to filter the data – it is the `where` condition in the SQL query that's being executed under the hood. Some of the other parameters we can use are as follows:

- `fields`: This parameter allows us to specify which attributes will be returned in the result. For example, if we only want the tutorial `title` field rather than the entire tutorial attribute, we can specify the title in the fields array.

- `start`: We can use this parameter to skip a certain number of entries. It is mainly used with **pagination**.

- `limit`: We can use this parameter to limit the number of entries that are returned.

- `sort`: This allows us to order the data based on a specific attribute. For example, we can use it to order the tutorials alphabetically or by the date they were created.

- `populate`: Used to indicate if we want to populate the relationship attribute.

Now that we have a service method ready, we can use it in the classroom controller. Let's update the classroom controller, as follows:

```
async findTutorials(ctx) {
    const {params} = ctx;
    const results = await strapi.service('
    api::classroom.classroom').findTutorials(params.id);
    return results;
}
```

We have updated the controller so that it returns the result from the `findTutorials` service method. If we save the changes and query our API, we will see that the `/api/classroom/1/tutorials` endpoint is returning a list of tutorials associated with the classroom:

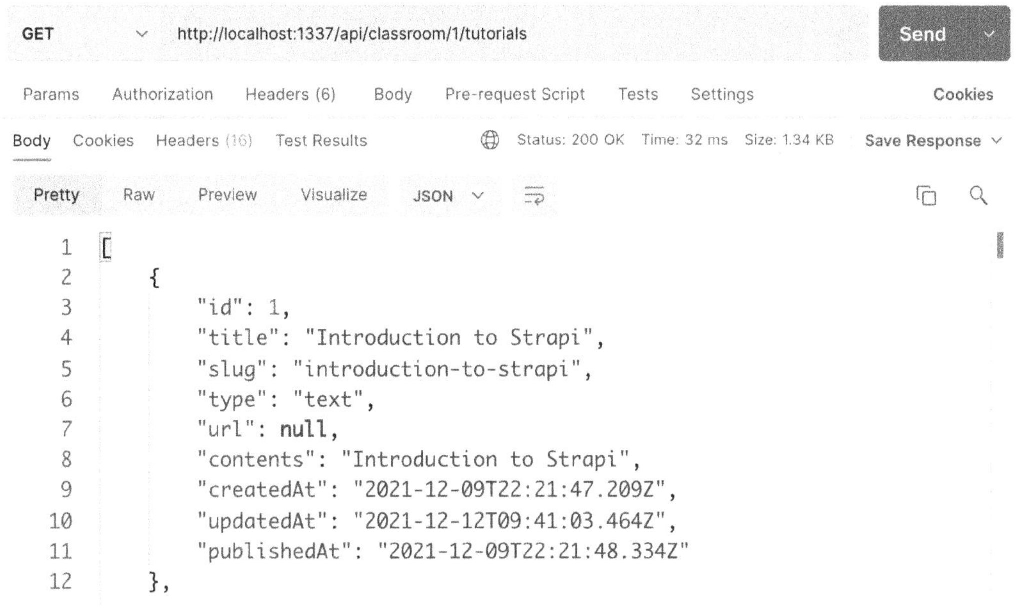

Figure 5.9: Get classroom tutorials API response

Returning a unified response

Looking at the response from the `/api/classroom/1/tutorials` endpoint, you will notice that the response structure is different from the rest of the API endpoint. All the other default endpoints in the API have a unified structure. Having a unified response structure is a great practice that helps in understanding the data. It also makes it easy to consume the data from a client later.

Strapi provides a transform utility that, as its name suggests, will transform the data into a unified format that follows the API response structure. This utility is defined in the **Strapi** `Core` controller, which means we can directly use it in any controller since all controllers inherit from the `Core` controller. Let's update the classroom controller and use the `transformResponse` utility:

```
async findTutorials(ctx) {
    const {params} = ctx;
    const results = await strapi.service('
    api::classroom.classroom').findTutorials(params.id);
    return this.transformResponse(results);
}
```

If we call the GET /api/classroom/1/tutorials endpoint one more time, we will see that the response follows the Strapi API response structure:

Figure 5.10: Get classroom tutorials API response after using transformResponse

Populating relationships

One thing you may have noticed in the /api/classroom/1/tutorials endpoint response is that there is no relationships information in the response payload. In other words, we cannot see the coverImage or classroom relationships. This is because, by default, Strapi will not populate those relationships unless we specify that we want to populate those relationships. So, let's get started.

As you may recall from the previous section, one of the parameters that we passed to entityService.findMany was the populate parameter, which we can use to indicate if we want to populate the relationship attribute. There are several ways to specify which relationships we want to populate; we can use an array of the names content-type to populate those relationships or use asterisk * to populate everything. If we want more control over the populated fields, such as by specifying which attribute for what or its sort order, we can use an object. For our case, let's do the following:

1. Let's update the populate parameter by passing an array of the content-types we want to populate, namely coverImage or classroom. The updated classroom service code should look as follows:

```
module.exports = createCoreService(
'api::classroom.classroom',
({ strapi }) => ({
  findTutorials(classroomId) {
  return strapi.entityService.findMany('
api::tutorial.tutorial', {
    filters: { classroom: classroomId },
    Populate: ['classroom','coverImage']
  });
  },
})
);
```

2. Let's test it out in Postman by sending GET /api/classroom/1/tutorials. We should see that the classroom relationship has been populated in the response:

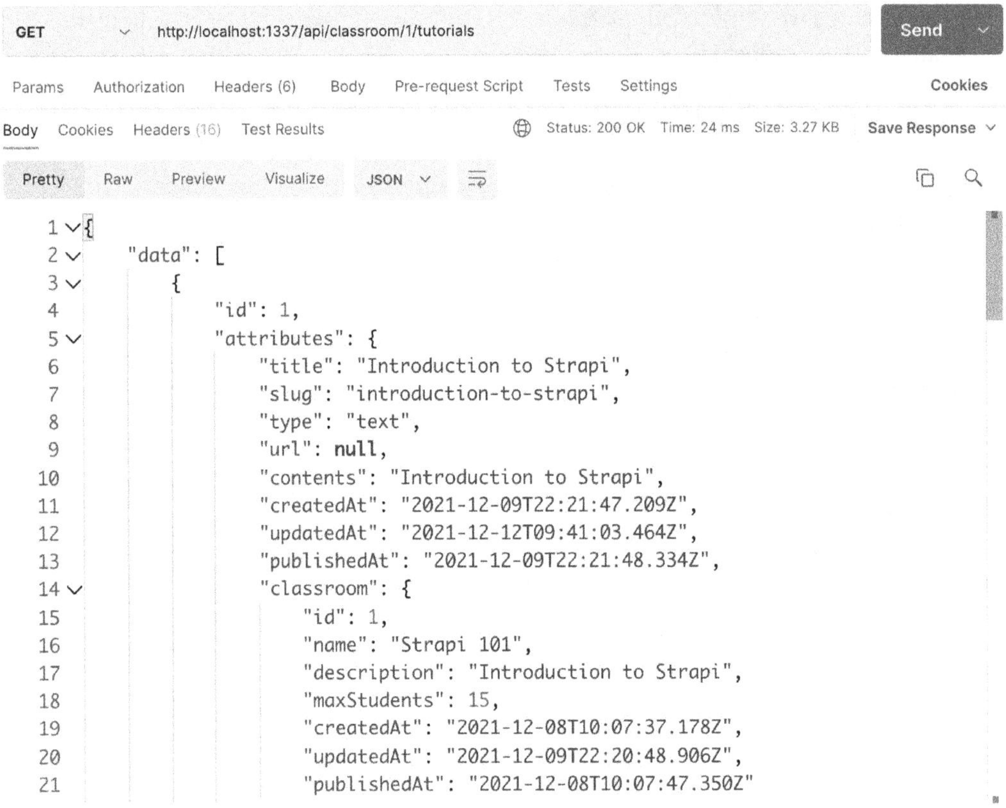

```
GET          ∨   http://localhost:1337/api/classroom/1/tutorials                    Send  ∨

Params   Authorization   Headers (6)   Body   Pre-request Script   Tests   Settings              Cookies

Body  Cookies  Headers (16)  Test Results          Status: 200 OK  Time: 24 ms  Size: 3.27 KB   Save Response ∨

Pretty   Raw   Preview   Visualize   JSON ∨  ⇥                                        ⎘  🔍

 1 ∨ {
 2 ∨     "data": [
 3 ∨         {
 4               "id": 1,
 5 ∨             "attributes": {
 6                   "title": "Introduction to Strapi",
 7                   "slug": "introduction-to-strapi",
 8                   "type": "text",
 9                   "url": null,
10                   "contents": "Introduction to Strapi",
11                   "createdAt": "2021-12-09T22:21:47.209Z",
12                   "updatedAt": "2021-12-12T09:41:03.464Z",
13                   "publishedAt": "2021-12-09T22:21:48.334Z",
14 ∨                 "classroom": {
15                       "id": 1,
16                       "name": "Strapi 101",
17                       "description": "Introduction to Strapi",
18                       "maxStudents": 15,
19                       "createdAt": "2021-12-08T10:07:37.178Z",
20                       "updatedAt": "2021-12-09T22:20:48.906Z",
21                       "publishedAt": "2021-12-08T10:07:47.350Z"
```

Figure 5.11: API response and its relationship populated

This works great for one-level relationships, but what about deep-level relationships? In other words, if classroom has a relationship with another content-type, how can we include that in the API response?

3. To illustrate this point, let's add a manager field to the classroom content-type. This will be a relationship to the users-permissions table (not Users as these are the Strapi admin users) that we'll consider the manager of that specific classroom.

Let's go back to the Strapi admin panel (it's been a while!), navigate to the content-type builder, and add that relationship:

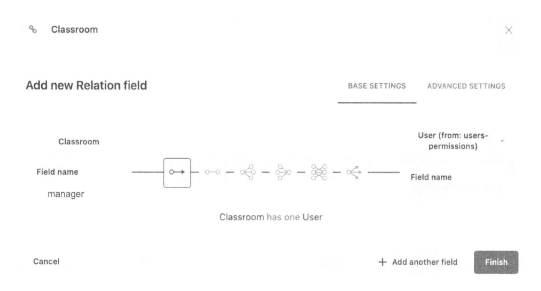

Figure 5.12: Adding a manager relationship to the Classroom content-type

4. Now, let's go to **Users** (under **Collection Types**) and add a new user:

Figure 5.13: Adding a new user under the Users collection type

Note that this is a normal API user, not a Strapi admin user. We touched on the differences briefly in the previous chapter and we will discuss the difference in more detail in *Chapter 7, Authentication and Authorization in Strapi*. But for now, we can think of those as the external users of our API (not the internal admins who have access to Strapi's administration).

5. Once we have added the user, we can assign that user as a manager of the classroom and save the entity:

Figure 5.14: Assigning our new user as a manager of the classroom

6. Now, if we call the `/api/classroom/1/tutorials` endpoint, we will notice that the manager field is not populated. This is because we only specified the classroom in the `entityService.findMany` parameter. Let's update it so that it also includes the classroom manager. The updated service code is illustrated here:

```
module.exports = createCoreService(
'api::classroom.classroom',
({ strapi }) => ({
  findTutorials(classroomId) {
  return strapi.entityService.findMany('
api::tutorial.tutorial', {
     filters: { classroom: classroomId },
     Populate: ['classroom','classroom.manager',
     'coverImage']
  });
  },
```

```
    })
  );
```

Now, our response will include the classroom and its related entity manager, along with its details:

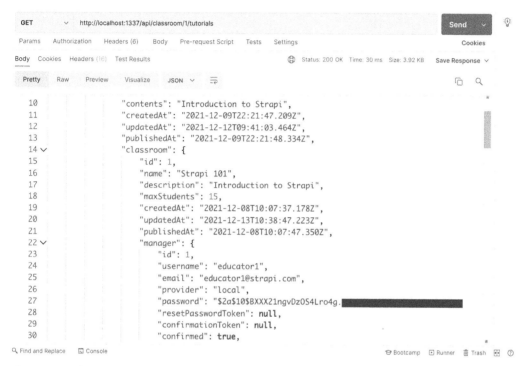

Figure 5.15: Our API now returns relationships two levels deep, the classroom entity, and the manager entity related to it

The response contains `Tutorials` assigned to `Classroom`. This is what we wanted, but there are still a few bits to improve. For example, the response contains the full user object for the `manager` field. This includes the **hashed** password of the user. These fields are marked as private in Strapi but since we used the **Entity Service API** directly, these options were not respected:

```
"manager": {
    "id": 1,
    "username": "educator1",
    "email": "educator1@strapi.com",
    "provider": "local",
    "password": "$2a$10$BXXX21ngvDzOS4Lro4g.▇▇▇▇▇▇▇▇▇
    "resetPasswordToken": null,
    "confirmationToken": null,
    "confirmed": true,
    "blocked": false,
    "createdAt": "2021-12-12T07:43:43.832Z",
    "updatedAt": "2021-12-12T07:43:43.832Z"
}
```

Figure 5.16: The API response includes all the properties of the object – even sensitive information, such as the hashed password

> **What is Private Data in Strapi?**
>
> Strapi allows us to mark any field in any content-type as private. This means that they will not be returned in API responses. This is useful for any field that we want to have a record of internally, but we don't want to expose it in the API response. Those fields are removed from the API response when we use the Strapi sanitize utility.

Sanitizing data – hiding passwords

To avoid such a potential leak from our API, we should always **sanitize** our responses. This is done by default by the `Core` controllers provided by Strapi, but if we customize those or create new controllers and services, then it is our responsibility to do this data sanitization. Luckily, Strapi provides us with a handy method to perform this operation. Let's update the classroom controller and use the `sanitize` utility:

```
"use strict";

/**
 * classroom controller
```

```
*/
const { createCoreController } = require('@strapi/strapi').
factories;
const { sanitize} = require('@strapi/utils');


module.exports = createCoreController ('api::classroom.
classroom', ({ strapi}) => ({
  async findTutorials(ctx) {
    const { params } = ctx;
    const results = await strapi.service('
    api::classroom.classroom').findTutorials(params.id);

    const model = strapi.getModel('
    api::tutorial.tutorial');
    const sanatizedResults = await
    sanitize.contentAPI.output(results, model);

    return this.transformResponse(sanatizedResults);
  },
});
```

The `sanitize.contentAPI.output` method expects two parameters. The first is the data we want to sanitize, while the second is the content-type schema of the data. We can get the schema by calling the `getModel` helper and passing to it the `uid` property of the content-type we want to use.

Now that we have applied the `sanitize` method to the response that was returned by the controller, we will no longer get private fields, such as **password** fields, in our response.

At this point, our new endpoint is working as expected: we added a new route and connected it to a controller handler that uses a custom service to perform the operation we wanted. However, in some scenarios, we don't want to add a new route – we're only interested in changing the behavior of an existing CRUD endpoint.

Overriding the default CRUD APIs

Out of the box, when we created our content-type, Strapi generated blank controllers and services. Even though they are blank, they do function (since our API is returning a response). For instance, the GET /classrooms endpoint maps to a handler called classroom.find. This is part of the Core controllers that Strapi gives us.

Let's imagine we have the following requirement:

As a dodgy company

I want to exaggerate the count of tutorials I am offering when displaying the classroom list

So that we give the impression we have a lot more to offer

For us to satisfy this requirement, we will need to override the default classroom.find controller action. As we mentioned earlier in this chapter, Strapi uses a unified structure for the API requests and responses. The API response consists of data and meta objects.

We will modify the meta object by adding a fake totalTutorials property and leaving the default find implementation as-is. To do so, we can update the controller like so:

```
async find(ctx) {
  // Calling the default core action
  const { data, meta } = await super.find(ctx);
  meta.totalTutorials = data.length + 100;
  return { data, meta };
},
```

Since we do not want to alter the `find` action itself, we used `super.find` to use the default `Core` controller implementation. Then, we added a new dodgy field to the metadata object by counting the total classrooms, adding an extra 100 to it, and calling that field `totalTutorials`. The point we want to illustrate here, regardless of the actual scenario we have implemented, is that we're able to change the implementation of any of the default API calls completely if we'd like. As always, Strapi provided a good starting point but it didn't lock us into doing anything.

With that, we have created new routes with new controller handlers and services, and we have also learned how to modify the behavior of the default CRUD API endpoints. These are the main components we need to customize an API and make it work the way we want. Sometimes, though we want more – we want to perform extra tasks at the point of querying the database, which can include manipulating a query, hiding a result, notifying another system, or any other requirement. In the next section, we will dig a level deeper into the database and explore a powerful mechanism that we can use to customize the API's behavior in Strapi: model life cycle hooks.

Tweaking database queries and responses with life cycle hooks

A model is a representation of a database entity. An entity in a system has a life cycle: it gets created, updated, deleted, and retrieved. While the routes are the external interface to the outside world, the database is the innermost level of our system. Strapi communicates with the database using models that represent it in a database-agnostic way (that is, it can work with different database types) and using the **Query Engine API** (`strapi.db.query`).

We briefly mentioned the Query Engine API earlier in this chapter when we came across the Entity Service API and learned that the Entity Service API uses the Query Engine API under the hood to execute database queries. The Query Engine API provides the glue between our models and the actual database entity; it transforms our requests into the final queries to be run against the database and then maps the result back into a model. Sometimes, we want to do something extra before or after running the queries, and this is where **life cycle hooks** come into play:

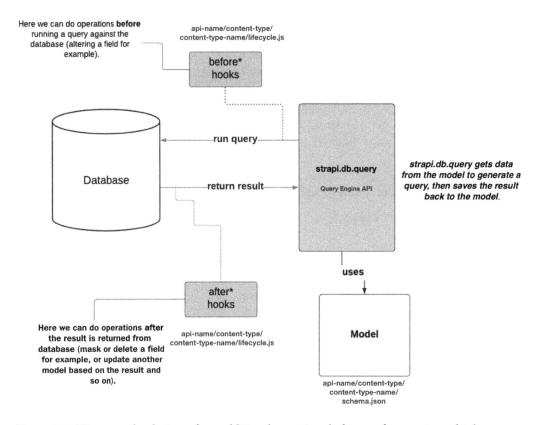

Figure 5.17: We can use hooks to perform additional operations before or after running a database query

Strapi provides **life cycle hooks**, where we can define `hooks` to perform additional operations before or after any database event. These hooks can be useful in the following scenarios:

- To delete, override, or mask certain fields before returning them to the user
- To automatically update a field based on other fields before creating the entity or updating it
- To notify other systems about a certain event

These hooks are as close as we'd get to our database in our Strapi code. So, while we could have such logic in the services or controllers, there is no guarantee that another endpoint will go through this service. They will all go through the life cycle hooks, though, so long as we are using `strapi.db.query` methods, as we should. Hooks will also run when we interact with an entity from the admin panel. The calls from the admin panel don't go through our API (and its services and controllers) but they still go through the model and its life cycle hooks, so any code that's added as a hook is guaranteed to run.

The before and after hooks that are provided are as follows:

- `beforeFindMany, afterFindMany`
- `beforeFindOne, AfterFindOne`
- `beforeCreate, afterCreate`
- `beforeCreateMany, afterCreateMany`
- `beforeUpdate, afterUpdate`
- `beforeUpdateMany, afterUpdateMany`
- `beforeDelete, afterDelete`
- `beforeDeleteMany, afterDeleteMany`
- `beforeCount, afterCount`
- `beforeCoundMany, afterCountMany`

These hooks live in `src/api/{api-name}/content-types/content-types/{content-type-name}/lifecycle.js`. By default, this file does not exist, so we will need to create it if we want to use these life cycle hooks. Let's do that now by creating the `lifecycles.js` file in `src/api/tutorial/content-types/tutorial/lifecycles.js`.

Next, let's make use of hooks to understand them better.

Using hooks

We have a new requirement for our API: we want to add a new field to the `tutorial` content-type called `summary`:

As a class manager

I want to generate a summary of the tutorial

So that it can be displayed to potential students

The field will be automatically generated based on the first 200 characters of the `contents` field.

Let's add the summary field to the tutorial content-type:

1. First, let's add a long text field called `summary` to the Tutorial content-type with a field. Let's also put a constraint on it to be `200` characters maximum. Add these lines to the model definition in `src/api/tutorial/content-types/tutorial/schema.json`, under the `attributes` object, or use **Content-Types Builder** in the admin panel to update the file automatically by adding the following lines:

    ```
    "summary": {
            "type": "text",
            "maxLength": 200
    },
    ```

2. Now, we can use the `beforeCreate` hook to generate the summary based on the contents field when creating a new tutorial:

    ```
    module.exports = {
      beforeCreate(event) {
        const { params: { data }  } = event;
        if (data.contents) {
            data.summary = data.contents.substring(0,
            200);
        }
      },
    };
    ```

The `beforeCreate` hook we defined here receives an `event` object. The `event` object contains several interesting properties, one of which is a `params` property. The `data` property that's being created is in the `params` object; we can mutate the data in whichever way we want before it is saved. So, here, we check if it has `contents` set, and then take the first `200` characters of it and assign them to `summary`.

3. Now, head to **Content Manager** in the admin panel and add a new tutorial entity with a `contents` field (and leave **Summary** empty). Then, save the entity. We will see that the **Summary** field is automatically generated:

← BACK

LifeCycle Hooks
API ID : tutorial

Title*

LifeCycle Hooks

Slug*

Type

text

Url

Contents

A model is a representation of a database entity. An entity in a system has a lifecycle: it gets created, updated, deleted, and retrieved. While the routes are the external interface to the outside world, the database is the most inner level of our system. Strapi communicates with the database

CoverImage

Click to select an asset

Summary

A model is a representation of a database entity. An entity in a system has a lifecycle: it gets created, updated, deleted, and retrieved. While the routes are the external interface to the outside wo

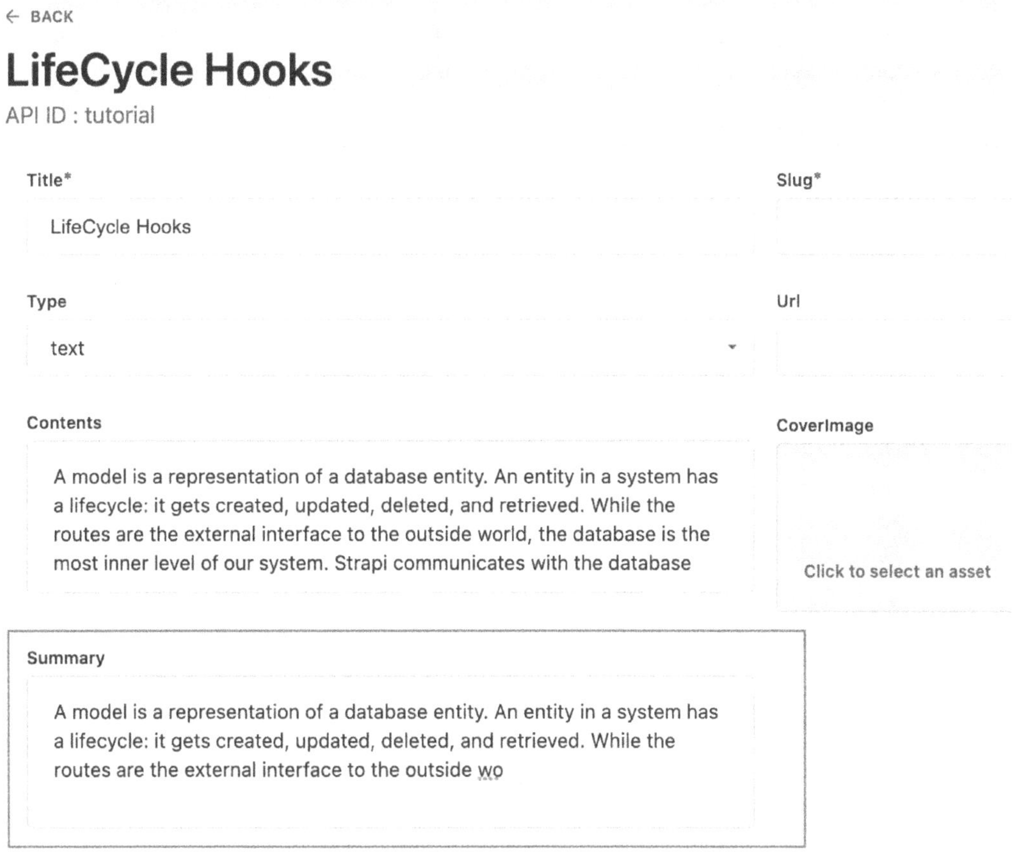

Figure 5.18: The summary was automatically generated by the beforeCreate hook based on the contents field

Now, you may have noticed that when we updated the contents field, the summary field does not get updated. That's because we only added a `beforeCreate` hook, not a `beforeUpdate` hook. We can duplicate the same method for `beforeUpdate`, but then we'd be repeating code. We can extract the method to put the summary in a function and reuse it, but a better solution would be to make use of Strapi **services** to encapsulate and reuse that logic.

DRYing the hook's logic with services

To make the code of the hook follow the **DRY** principle, we will encapsulate the logic of generating the summary in a service. For now, we will make it part of the tutorial service, though we could also create a service independent of any content-type. Let's get started:

1. To create the service method, we must edit the `tutorial service` module in `src/api/tutorial/services/tutorial.js` and add our helper method there:

    ```
    'use strict';
    /**
     * tutorial service.
     */
    const { createCoreService } = require('@strapi/strapi').
    factories;
    module.exports = createCoreService('api::tutorial.
    tutorial', () => ({
      generateSummary(data) {
        if (data.contents) {
          return data.contents.substring(0, 200);
        }
          return null;
        },
    }));
    ```

2. Now that we have added `generateSummary` to the service, we can use it anywhere in our Strapi API by using `strapi.services('api::tutorial.tutorial').generateSummary`. So, let's update our life cycle hook so that it can use it:

    ```
    module.exports = {
      beforeCreate(event) {
    ```

```
      const { params: { data } } = event;
      data.summary = strapi.service('
      api::tutorial.tutorial').generateSummary(data);
  },
  beforeUpdate(event) {
      const { params: { data } } = event;
      data.summary = strapi.service('
      api::tutorial.tutorial').generateSummary(data);
  },

};
```

Now, we're calling our service method from both hooks without duplicating the code and making use of the concept of services from Strapi. We also didn't have to import this method explicitly as Strapi makes all the service methods available so that they can be used in other components.

Note

To mutate or not to mutate

The life cycle hooks normally receive parameter data that we can mutate to alter the data we're about to pass to or from the database query. This is how Strapi works at the moment. For the service logic, we chose not to mutate the data object directly. This makes the code easier to understand, predict, and test, as we shall see in *Chapter 11, Testing the Strapi API*.

Under the hood – the scope of services

When we start the Strapi server, a global `strapi` object is created and then *bootstrapped*. The bootstrapping process includes looping through all our models, controllers, services, and routes (and other components) and attaching them to the global `strapi` object. In this process, the Core API is also created, which creates the default services and controllers – the magic ones that handle CRUD operations by default, even though we didn't write any code. Since the `strapi` object is global and accessible from anywhere, we can access any service with `strapi.services(uid).methodName`.

We now have two functional life cycle hooks – one is triggered before creating a content-type, while the other is triggered before updating the content-type. We have encapsulated the actual login into the tutorial service to help us have clean code without repeating ourselves.

Summary

In this chapter, we did a lot of coding. Previously, we mainly relied on the admin panel or the CLI to generate the API for us. This is amazing and one of the reasons we love Strapi – we can build 80% of our system with a few clicks on a slick UI. This chapter was about the 20% of use cases when the default is not enough, and when we want to customize the API to do a bit more or a bit less.

We dug deeper into how an API is built in Strapi by exploring the concepts of routes, controllers, services, models, and the model life cycle. We learned how to add new routes and how to map them to controllers that make use of services. We also explored some security considerations, such as sanitizing sensitive data. We learned how to disable default core endpoints, which can come in handy when we don't want to expose a certain operation.

We also looked at how to customize the API by using the model life cycle hooks, which provide us with a level of integration that's close to the database where we can run logic before or after running database queries.

In the next chapter, we will look at the Content API and all the options Strapi gives us by default to filter, sort, paginate, and interact with our APIs.

6
Dealing with Content

In this chapter, we will see how we can interact with API content. Strapi provides a handy set of API parameters that can be used to perform common tasks such as filtering, sorting, and paginating content. We will see how can we use those API parameters to manipulate API results. Finally, we will explain how we can use the Strapi Query Engine API to write and execute complex query operations.

The topics we will cover in this chapter are as follows:

- Creating sample data
- Sorting and pagination with API operators
- Filtering content with Strapi API filter operators
- Paginating API content
- Understanding how Strapi works under the hood

Creating sample data

Before we get core of this chapter, we will need sample data to interact with. We created a couple of entries in our API in the previous chapter; however, since this chapter is all about content filtering and sorting, we will need more than a couple of entries in the database.

In *Chapter 5, Customizing Our API*, we saw how we can add new endpoints to our API. Let's use the same strategy to add a temporary API endpoint to populate some data in our database; we will use this API endpoint to create a few classrooms in our API.

> **Is This the Best Way to Get Sample data?**
>
> Adding a new API endpoint here is just a quick and dirty way to populate data in the database; however, this is not the recommended way to seed data into the system. We will cover the proper way to do so in *Chapter 9, Production-Ready Applications*. We're using the endpoint method, as we should be familiar by now with customizing Strapi routes and controllers; also, introducing the seed mechanism at this stage might be a bit of a distraction.

Let's get started:

1. Since we are creating a few classrooms, we will modify the `src/api/classroom/controllers/classroom.js` classroom controller, adding a new `seed` method:

```js
async seed(ctx) {
  try {
    const classroomsPromise = [];
    // Min and Max values to generate random number in
       range
    const min = 1;
    const max = 30;
    // Number of classrooms to be created
    const numberOfClasses = 50;

    Array(numberOfClasses)
      .fill(null)
      .forEach((_item, index) => {
        const name = `classroom_${index + 1}`;
        // Get random numnber in range of min and max
        const maxStudents = Math.random() * (max - min +
        1) + min;
```

```
        classroomsPromise.push(
          strapi.services(
           'api::classroom.classroom').create({
            data: {
              name,
              description: `Description of the classroom
              ${name}`,
              maxStudents: Math.floor(maxStudents),
            }
          })
        );
      });

      await Promise.all(classroomsPromise);
      return { message: "Ok" };
    } catch (e) {
      strapi.log.error("Failed to seed the database");
    }
  },
```

The seed method is a simple method that will generate 50 classrooms with names ranging from classroom_1 to classroom_50, and a random maxStudents number.

2. Now that we have the API endpoint code implemented, let's update the src/api/classroom/routes/custom-classroom.js router file to add this endpoint to our API routes. At the beginning of the routes array, add the following definition:

```
{
    "method": "POST",
    "path": "/classrooms/seed",
    "handler": "classroom.seed",
    "config": {
        "auth": false
    },
},
```

This change will add a POST /api/classrooms/seed endpoint to our API that we can use to generate sample data to work with. Let's try it in Postman by sending a POST request to the newly added endpoint. We should see a response with an Ok message:

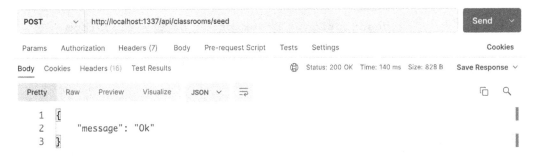

Figure 6.1: The /classrooms/seed endpoint response

To confirm we have the data, let's issue a GET /api/classrooms request against our API. We should be able to see at least 50 entries in the response.

Cleaning up

Once we have the data in our database, we do not need the /api/classrooms/seed endpoint anymore. We should remove it, along with the route created in the src/api/ classroom/routes/custom-classroom.js file.

Now that we have our database seeded with sample data, we are ready to interact with API content. Next, let's see how can we sort the data according to a specific field.

Sorting API content

As we have seen so far, Strapi provides us with defaults that allow us as developers to speed up some of the common tasks with minimal effort. A common task when dealing with API content is sorting data according to a specific field.

To sort the data, we can use the sort parameter followed by the field we want to sort on and the direction of the sort. The general format is ?sort=FIELD_NAME:[ASC OR DESC].

To illustrate how the sort parameter works, let's assume we have the following requirement:

As an API user
I want to be able to sort the classrooms by the maximum number of students from the largest to the smallest

To satisfy this requirement, we will use the `sort` API parameter on `maxStudents`; the API URL will be `GET /api/classrooms?sort=maxStudents:desc`. Let's test it with Postman; we should see the classrooms sorted by the highest `maxStudents` number first:

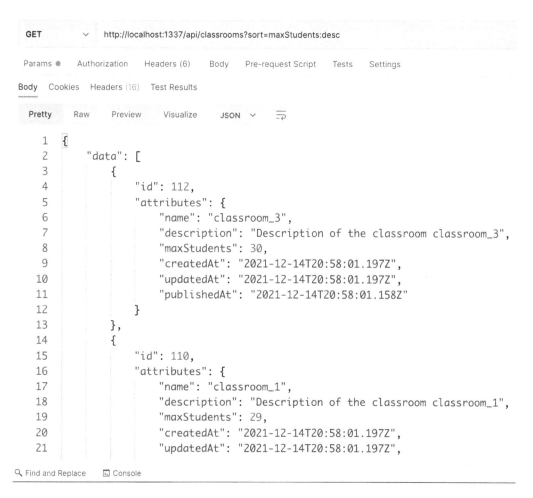

Figure 6.2: An example of sorting classrooms by the number of maximum students

> **Important Note**
>
> You will see slightly different data from the preceding screenshot as the `maxStudents` number is randomized. However, the classrooms should be ordered using `maxStudents` from largest to smallest.

What about sorting on multiple fields?

What if our requirements changed to sorting the data on multiple fields? Let's assume we have the following new requirement:

As an API user
I want to be able to sort the classrooms by most recent first and then by the lowest number of maximum students

This new requirement means we need to sort the results by the `createdAt` field in descending order and by `maxStudents` in ascending order. Luckily, the `sort` API parameter has got us covered; we can use a comma to indicate that we want to sort on multiple fields. The general format when sorting on multiple fields is as follows: `?sort=FIELD1:[ASC OR DESC],FIELD2:[ASC OR DESC]`.

To satisfy our new requirement, we will format our request URL to be `GET /api/classrooms?sort=createdAt:desc,maxStudents:asc`.

Do We Need to Pass :asc All the Time in the URL?

The default order to sort data is in ascending order, which means we can omit `:asc` from the URL. However, we're choosing to keep it here to make the URL as readable and explicit as possible.

Let's test the new URL with Postman and note the result:

Figure 6.3: An example of sorting classrooms by date and number of students

If you examine the response, you can note that the classrooms are sorted by the `createdAat` field first and then by the maximum number of students.

Sorting the data according to a specific field is one of the most common tasks when dealing with API content. Another frequent task is filtering the data, rather than having all the data available. In the next section, we will discuss how we can use the Strapi `filter` parameter to filter the API content.

Filtering API content

Filtering API content allows the API consumer to retrieve a subset of data rather than retrieving an entire dataset. Filtering only works with the endpoints that retrieve the entire data for a specific **content-type**; to be more specific, it will only work with the `find` endpoint.

To filter the API data, we can use the `filters` API parameter. Strapi uses **LHS bracket** syntax for query parameters, which means the query parameters are encoded in the URL using square brackets (`[]`). The `filters` API parameter is in the `?filters[FILED_NAME][OPERATOR]=VALUE` format, where `FILED_NAME` is the name of the field we want to filter on and `OPERATOR` is the operator we want to use to filter the data. The following table shows all the possible operators we can use:

Filter operator	Description
`$eq`	Equal to
`$ne`	Not equal
`$containsi`	Contains
`$contains`	Contains case sensitive
`$ncontainsi`	Does not contain
`$ncontains`	Does not contain case sensitive
`$lt`	Less than
`$lte`	Less than or equal to
`$gt`	Greater than
`$gte`	Greater than or equal to
`$in`	Included in an array
`$nin`	Not included in an array
`$null`	Is null
`$notNull`	Is not null
`$between`	Value between
`$startsWith`	Starts with
`$endsWith`	Ends with

Table 6.1 – A table showing the usable filter suffix values

To better illustrate the `filters` parameter, let's consider the following requirement:

As an API user
I want to be able to retrieve all classrooms that include the classroom_3 string in the name

To achieve this requirement, we can use the $contains or $containsi filters; our request URL would be GET /api/classrooms?filters[name] [$contains]=classroom_3. If we called our API with this parameter, we would get 11 results back, classroom_3 and classroom_30 to classroom_39:

Figure 6.4: An example of using "contains" filter suffixes

The previous two examples are somewhat simple, so the URL is easily formatted and can be done manually. But what if we had a slightly more complex requirement?

For such a situation, it is recommended to use a query string parsing library. Strapi recommends using the qs library. qs is a popular query string parsing and stringifying library, which is used internally by Strapi and is recommended for parsing nested objects to create complex queries.

To demonstrate, let's assume we have the following new requirement:

As an API user
I want to be able to retrieve all classrooms that include the classroom_3 string in the name
AND where the maximum number of students is less than 10 OR greater than 25

Let's analyze this requirement: we already did the first part by filtering by a string included in the name, so we will use the $contains operator with the name field. For the maximum number of students, we will need to use the $lt and $gt operators on the maxStudents field to get the results of less than and greater than respectively.

This query is complex to write manually; instead, let's use the qs library, as suggested by Strapi:

1. At the root of the project, create a new file called query_helper.js.

2. Add the following code:

```
const qs = require("qs");
// name contains classroom_3 AND number of max students
is greater than 25 OR less than 10
const query = qs.stringify(
  {
  filters: {
    $or: [
      {
        maxStudents: {
          $lt: 10,
        },
      },
      {
        maxStudents: {
          $gt: 25,
        },
      },
    ],
    name: {
      $contains: 'classroom_3',
    },
  },
},
```

```
    {
        encodeValuesOnly: true, // prettify url
    }
);
console.log(query);
console.log(`\nhttp://localhost:1337/api/
classrooms?${query}\n`);
```

3. Save the changes to the file.

In the first line of the previous code snippet, we require the qs library. Since we created query_helper.js in the root of our project, we do not need to install the library, as it's included as a dependency for Strapi.

Next, we used the stringify method to create a string for the query. Then, we define our filters and their operators. Note that we did not specify anything in the filters array to indicate an AND operation but we explicitly used $or for the maxStudents query; this is because, by default, different operators are joined using an implicit AND operation.

Finally, we print out the raw query itself, as well as a handy link to the classroom endpoint. To run this file from the command line, execute the node query_helper.js command:

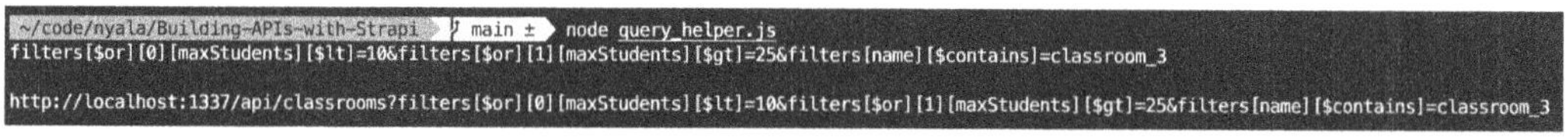

```
~/code/nyala/Building-APIs-with-Strapi  ⑂ main ±  node query_helper.js
filters[$or][0][maxStudents][$lt]=10&filters[$or][1][maxStudents][$gt]=25&filters[name][$contains]=classroom_3

http://localhost:1337/api/classrooms?filters[$or][0][maxStudents][$lt]=10&filters[$or][1][maxStudents][$gt]=25&filters[name][$contains]=classroom_3
```

Figure 6.5: Running the query_helper file

Let's copy the full URL and test it in Postman. Copy the second line as is and paste it into Postman:

```
GET         http://localhost:1337/api/classrooms?filters[$or][0][maxStudents][$lt]=10&filters[$or][1][maxStudents][$gt]=25&filters[name][$contains]=classroom_3

Params ●    Authorization    Headers (6)    Body    Pre-request Script    Tests    Settings

Body    Cookies    Headers (16)    Test Results                              Status: 200 OK    Time: 23 ms

Pretty    Raw    Preview    Visualize    JSON ∨

1  {
2      "data": [
3          {
4              "id": 112,
5              "attributes": {
6                  "name": "classroom_3",
7                  "description": "Description of the classroom classroom_3",
8                  "maxStudents": 30,
9                  "createdAt": "2021-12-14T20:58:01.197Z",
10                 "updatedAt": "2021-12-14T20:58:01.197Z",
11                 "publishedAt": "2021-12-14T20:58:01.158Z"
12             }
13         },
14         {
15             "id": 140,
16             "attributes": {
17                 "name": "classroom_31",
18                 "description": "Description of the classroom classroom_31",
19                 "maxStudents": 9,
20                 "createdAt": "2021-12-14T20:58:01.198Z",
21                 "updatedAt": "2021-12-14T20:58:01.198Z",

Q Find and Replace    Console
```

Figure 6.6: A complex query result

If you examine the result, you will see that only classrooms with the classroom_3 string and a maxStudents number of either less than 10 or greater than 25 are included in the response.

We are able to sort and filter API data. In the next section, we will discuss another important operation when building an API, the ability to limit and skip certain content to achieve API pagination.

Paginating API content

In the previous two sections, we were able to filter and sort API data, but we did not have control over how many records were being retrieved. In some situations, where we have a large number of records, we will want to introduce a boundary or limit to the number of records being retrieved by a single API call. Luckily, Strapi has us covered as well for such situations: it provides a pagination API parameter that we can use to limit or paginate the API results.

> **Important Note**
>
> Similar to the `filters` API parameter, the `pagination` API parameter can only be used with the `find` endpoint.

The pagination metadata is included in the API response under the `meta` object. By default, Strapi returns *25 records per page*. To change this value, we can use the `pagination[pageSize]` parameter, which will allow us to control how many records we want to retrieve per page.

To better illustrate this, let's assume we have the following requirement:

As an API user
I want to be able to restrict the number of classrooms retrieved by an API call to 10 records per page, and I want to be able to move between pages

Using the `pagination[pageSize]` API parameter, we can construct our API URL with `GET /api/classrooms?pagination[pageSize]=10`. If we test this URL in Postman, we will only get 10 classrooms per page:

Figure 6.7: Using the _limit parameter in API calls

You can see that the `pagination` metadata is now showing 10 records per page and the page number is 1. To move to the next page or any other page, we can add the `pagination[page]` API parameter to the URL; the full URL would be GET /api/ classrooms?pagination[pageSize]=10&pagination[page]=2.

Another way to paginate the API result is to use the `limit` and `start` API parameters. These parameters allow us to use `offset` to paginate the data. To better illustrate these parameters, let's imagine that we have the following requirement:

As an API user
I want to be able to restrict the number of classrooms retrieved by an API call to 10 records only, skipping the first 15 records

To satisfy that requirement, we will use the following parameters to call the GET /api/ classrooms?pagination[limit]=10&pagination[start]=15 API:

Figure 6.8: Using the pagination limit and start parameters

> **Important Note**
>
> The `start` and `limit` parameters cannot be used with the `pageSize` and `page` API parameters at the same time.

Now that we are able to sort, filter, and limit API data, let's discuss in the next section how to do so by just using a few parameters in the URL.

Under the hood – how it works

You might be wondering how we can perform all of these operations just using API parameters. Does Strapi fetch all the data for a certain model from the database and then filter it according to the API parameters?

Under the hood, Strapi uses **Knex.js**, a popular multi-dialect SQL builder for Node.js. The API parameters are parsed using a query builder utility; the parsed result can then be used in Knex.js to run the database queries.

> **Important Note**
>
> If you are interested in checking out the filter query builder, you can find the source code in the `node_modules/@strapi/database/lib/query/query-builder.js` file.

In the previous chapter, we briefly mentioned the Query Engine API. The Query Engine API allows unrestricted access to the database layer at a lower level. It makes use of the query builder previously mentioned together with Knex.js to run and execute database queries. The Query Engine API is available through `strapi.db.query`.

Let's recall the following requirement from earlier in this chapter:

As an API user
I want to be able to retrieve all classrooms that include the classroom_3 string in the name
AND where the maximum number of students is less than 10 OR greater than 25

The same result can be achieved using `strapi.db.query`; for example, we can write the query as follows:

```
const results = await strapi.db.query('api::classroom.
classroom')
.findMany({
    where: {
      name: {
        $contains: 'classroom_3',
      },
      $or: [
        {
        maxStudents: {
          $lt: 10,
        },
        },
        {
        maxStudents: {
          $gt: 25,
        },
        },
      ],
      },
    orderBy: { id: 'desc' },
});
```

This query will return the exact same result as with the `filters` parameter. The Query Engine API is used mainly by plugin developers to add custom logic to their applications. So, as a rule of thumb, we should always use the core services and the Entity Service API when we are customizing an API and the Query Engine API when we are developing a Strapi plugin.

Summary

In this chapter, we saw how we can interact with API content and what operations Strapi gives us by default.

We started by creating a naïve seeder to populate the database with sample data to work with. Then, we had a look at sorting the data using the `sort` API parameter. Afterward, we explored data filtering using the `filters` API parameter and saw how it can be used to do simple and complex data filtering. Lastly, we used the `pagination` API parameter to paginate and limit the number of results returned by default and specify how many records we would like to skip.

Finally, we dug a bit deeper into the underlying structure of Strapi and saw how the database queries work.

In the next chapter, we will explore authentication and authorization in Strapi. We will see how we can allow signups and logins, and protect and secure some of the API endpoints.

7

Authentication and Authorization in Strapi

In this chapter, we will explore the topic of authentication and authorization in Strapi. We will start the chapter by explaining the difference between admin users and API users. Then, we will see how to work with login and sign-up functionality. Afterward, we will explain how to protect and secure certain API routes from unauthorized access.

The topics we will cover in this chapter are as follows:

- The difference between admin users and API users
- Login and sign-up in Strapi
- Securing API routes
- Roles and permissions
- Working with Strapi policies
- Overview of OAuth providers

By the end of this chapter, you will be able to implement authentication in the API, as well as adding a layer of authorization to prevent unauthorized users from accessing certain areas of the API.

Understanding the difference between admin users and API users

There are two types of users in Strapi: administrator panel users and API users. Mostly, we will be working with API users, but still, it is a good idea to know the difference between them.

Admin users are those who can log in to the Strapi admin panel. When we set up and installed Strapi in *Chapter 1*, *An Introduction to Strapi*, we created our first admin user, and we have been using that user ever since to interact and work with the admin panel. Additional admin users can be added from the **Settings** panel.

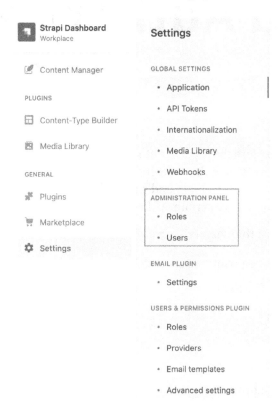

Figure 7.1: Admin user settings locations

The first admin users we create will have the role of **Super Admin**; they have access to manage all features and settings in the Strapi admin panel, but you cannot use them to interact with the API contents via the API endpoints.

API users, on the other hand, are considered another content-type in the API. Unlike the classroom and tutorial content-types, the user content-type is created automatically when we set up Strapi. The user content-type is managed via the core **users-permissions** plugin. We can use API users to interact with the API contents via the API endpoint; however, we cannot use an API user to log in to the Strapi admin panel.

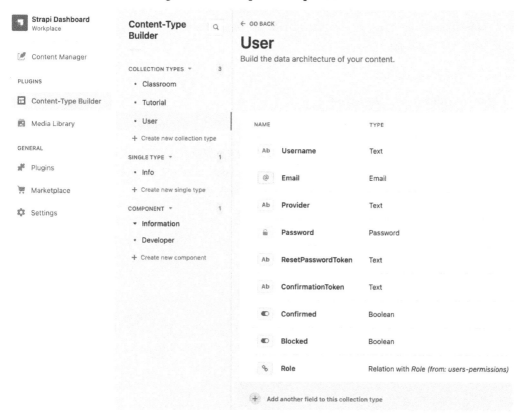

Figure 7.2: API users content-types

Now that we know and understand the difference between admin users and API users, let's see how can we log in and sign up to our API in the next section.

API login and sign-up

As mentioned in the previous section, users are managed via the **users-permissions** plugin. This plugin is not just there to manage users; it also adds an access layer to the API using a **JSON Web Token (JWT)** to authenticate users. Each time a request is made to a secure endpoint, a JWT must be present in the **Authorization** header. Additionally, the plugin exposes two API endpoints for sign-up and login. To better demonstrate this, let's consider the following requirement:

As an API user

I want to be able to register a new account with the following properties so that I can use the system using my identity:

- *Username: educator1*

- *Email: educator1@strapi.com*

- *Password: password*

To satisfy this requirement, we can simply use the `/api/auth/local/register` endpoint to create a new user account. In Postman, issue a `POST` request to `/api/auth/local/register` with the following as a JSON payload:

```
{
  "username": "educator1",
  "email": "educator1@strapi.com",
  "password": "password"
}
```

Strapi will create the user account and return a JWT as well as the user object details in the response. You can also see the user details from the Strapi admin panel under **Users collection type**.

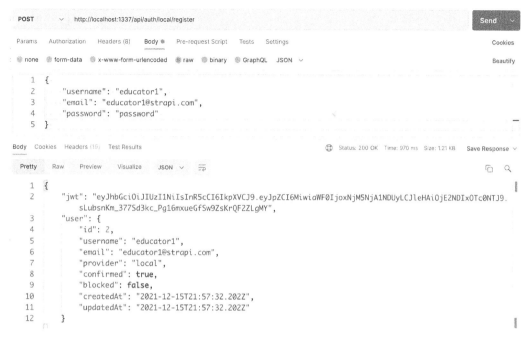

Figure 7.3: Registering a new user

Once the account has been created, the user can use it to authenticate themselves against the API. To illustrate this, let's consider this simple requirement:

As an API user

I want to be able to log in to my account

So that the system knows my identity

To satisfy this requirement, we can simply use the POST /api/auth/local endpoint to log in. This endpoint requires two parameters to be passed in the request body, identifier and password. The identifier can be either email or username. In Postman, issue a POST request to /api/auth/local. We will use email as an identifier but feel free to use username as well if you want to test it.

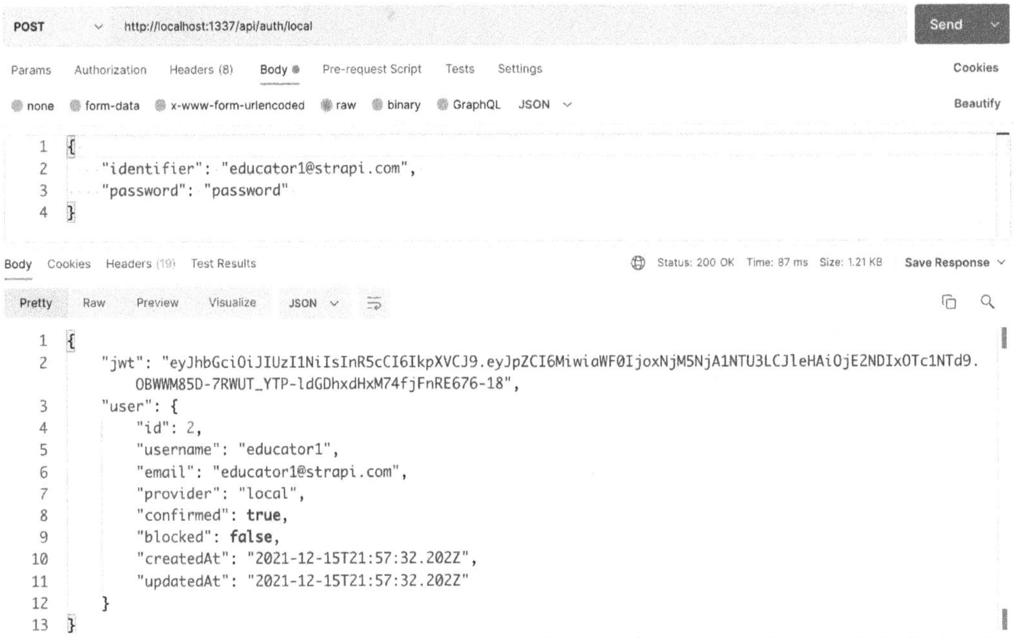

Figure 7.4: API login

If you pass the correct credentials, then you will get a JWT as well as the user object in the response body. If you examine the token using the https://jwt.io/ website, you can see that the token includes the user ID in the payload and the token will expire in *30 days*:

Encoded <small>PASTE A TOKEN HERE</small>

eyJhbGciOiJIUzI1NiIsInR5cCI6IkpXVCJ9.ey
JpZCI6MiwiaWF0IjoxNjM5NjA1NTU3LCJleHAiO
jE2NDIxOTc1NTd9.OBWWM85D-7RWUT_YTP-
ldGDhxdHxM74fjFnRE676-18

Decoded <small>EDIT THE PAYLOAD AND SECRET</small>

HEADER: ALGORITHM & TOKEN TYPE

```
{
  "alg": "HS256",
  "typ": "JWT"
}
```

PAYLOAD: DATA

```
{
  "id": 2,
  "iat": 1639605557,
  "exp": 1642197557
}
```

VERIFY SIGNATURE

```
HMACSHA256(
  base64UrlEncode(header) + "." +
  base64UrlEncode(payload),
  your-256-bit-secret
) ☐ secret base64 encoded
```

Figure 7.5: Decoded JWT

What if I want to change the JWT expiration time?

The 30-day token validity might not be suitable for all use cases; we might have a requirement that a token should be valid for 1 hour or 1 day. Generally speaking, as a best practice, a JWT should be short-lived. This is to ensure that even if a JWT were stolen or leaked, the impact would be minimal as the token would cease to be valid fairly quickly. To follow this best practice and reduce the security risk of exposing a JWT, we will change the JWT configuration. Strapi uses the popular jsonwebtoken module under the hood to manage and generate the JWT, and we can use the expiresIn option to adjust the token validity.

To better illustrate this, let's assume the following requirement:

As a security-concerned API user

I want my token to expire after 7 days

So that the security risk of exposing the token is reduced

To satisfy this requirement, we will need to create a new file called `plugins.json` in the config folder. The file's full path is `config/plugins.js`; we can then set the `expiresIn` option. The value of this option is either numeric, which will be assumed to be in seconds, or a string describing a time span. The time span string is parsed using the **vercel/ms** library (`https://github.com/vercel/ms`) where we can find the full documentation for expressing a time span (that is, 7d means 7 days):

```
Module.exports = env => ({
  "users-permissions": {
    "config": {
      "jwt": {
        "expiresIn": "7d"
      }
    }
  }
});
```

The following screenshot shows the newly created `plugins.js` file and its contents:

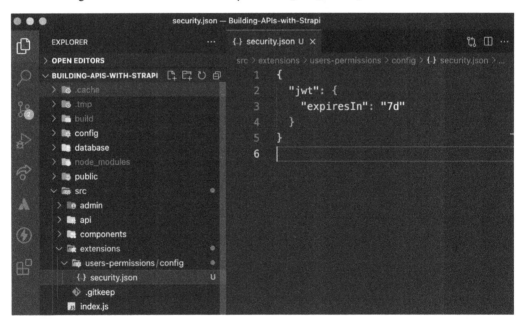

Figure 7.6: JWT expiration settings

Once you have created the file and saved the changes, you will need to start the development server again as it might have crashed while we were modifying the JWT settings.

> **Note**
>
> Unfortunately, Strapi might crash when updating certain settings. This will probably be fixed in upcoming versions but for now, the solution is to restart the server.

Once the server is up and running, you can issue a new JWT by logging in again using the `/api/auth/local` endpoint. Examine the JWT one more time; this time, you will see that it expires in 7 days rather than 30 days.

What if I want to change the JWT secret?

A JWT secret can be used to verify the token. By default, Strapi generates a token secret value and stores it in the `src/extensions/users-permissions/config/jwt.js` file. A new secret value can be provided using the `JWT_SECRET` environment variable. It is also possible to use different environment variable names by adjusting the value in the `src/extensions/users-permissions/config/jwt.js` file.

> **Updating JWT_SECRET**
>
> Using the auto-generated secret value is fine while we are working on the development environment; however, we should set a custom value using `JWT_SECRET` when deploying the API to a production environment. Otherwise, a malicious user with access to the secret would be able to create a token that impersonates a user and have the same access and permissions level as the impersonated user. We will cover this is in more detail in *Chapter 9, Production-Ready Applications*.

Having the login functionality and the ability to obtain a JWT will allow us to protect and secure some of our API endpoints. Let's discuss next how can we do that.

Securing API routes

All the API endpoints we have so far are publicly accessible. We will need to add a layer of security to our API so that only authorized users can perform certain actions, such as creating, editing, and deleting content.

Strapi makes it easier for us developers to add security to our API. By default, there are two roles in Strapi that we can use to manage permission and access. Those roles are as follows:

- **Public role**: This role is intended to be used by everyone to access the public endpoint of the system, for example, the get all classrooms or get tutorials endpoints.

- **Authenticated role**: This is the default role for all authenticated users. It is intended to manage access to protected areas of the API.

As we have seen so far, Strapi is very flexible and easily customized. We can alter those roles as we see fit, and we will do so later on in this chapter. But first, let's see how we can use the Authenticated role and protect certain API endpoints.

To better illustrate this, let's consider the following requirement:

As an owner of the API

I want only authenticated users to be able to create a new tutorial

So that the database is not filled with spam tutorials

If you remember, in *Chapter 3, Strapi Content-Types*, we made the create tutorial endpoint public. Let's revert that first. From the Strapi admin panel, do the following:

1. Click on **Settings** from the left-hand side menu.
2. Click **Roles** under **USERS & PERMISSIONS PLUGIN**.
3. Click **Public** to modify the permissions for the Public role.
4. Under **Tutorial** permissions, uncheck the permission for **create**.

Figure 7.7: Updating the create tutorial permission for the Public role

5. Save the changes and wait for the server to restart.

6. Open Postman and try to send a POST request to the /api/tutorials endpoint.

Figure 7.8: Access denied for creating tutorial

7. You will see a **Forbidden 403** response from the server.

Next, we will need to allow the Authenticated role to access the `create` tutorial endpoint:

1. Click on **Settings** from the left-hand side menu.

2. Click **Roles** under **USERS & PERMISSIONS PLUGIN**.

3. Click **Authenticated** to modify the permissions for the Authenticated role.

4. Under **Permissions**, check the permission for **create**.

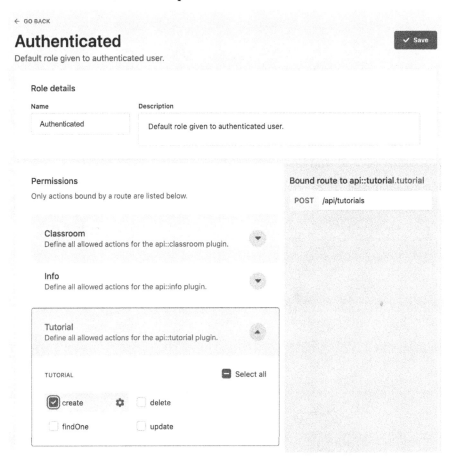

Figure 7.9: Authenticated role permission

5. Click the **Save** button to save the changes. Make sure to wait for the server to restart.

The create tutorial endpoint is now secure and only users with the Authenticated role can access it. To confirm this, let's try accessing the endpoint using the user we created earlier in this chapter:

1. If you do not have a JWT, log in to the API by sending POST /api/auth/local with the user information we created earlier.

2. Make note of the jwt value in the response.

3. Send a request again to the POST /api/tutorials endpoint but this time set the JWT in the Authorization header as **Bearer Token**.

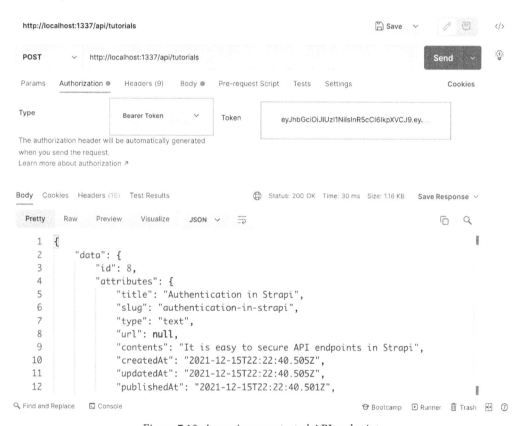

Figure 7.10: Accessing a protected API endpoint

4. Since we have a JWT in the Authorization header, the server will acknowledge the request and we will be allowed to access the create tutorial endpoint.

Under the hood – how it works

Each time a request is sent to the API, the server checks whether the `Authorization` header is present. If the header is not present, then permissions for the Public role will be used to verify whether access should be granted or not.

If there is an `Authorization` header, then the user ID will be extracted from the JWT, and access to the route will be granted or denied based on the user role.

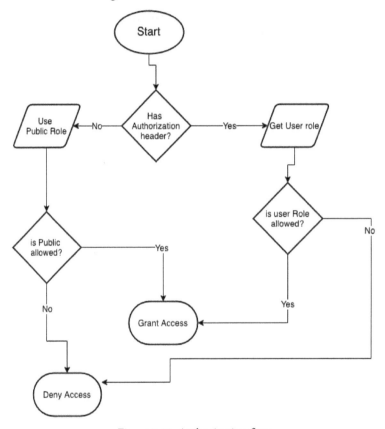

Figure 7.11: Authorization flow

Now that we understand how to secure our API endpoint, let's see how can we use the roles and permissions functionality in Strapi to grant users access to a specific resource in the API.

Using Strapi roles and permissions

Going back to the main actor use case diagram in *Chapter 2, Building Our First API*, the main actors in the API are **Students**, **Teachers**, and **Admins**.

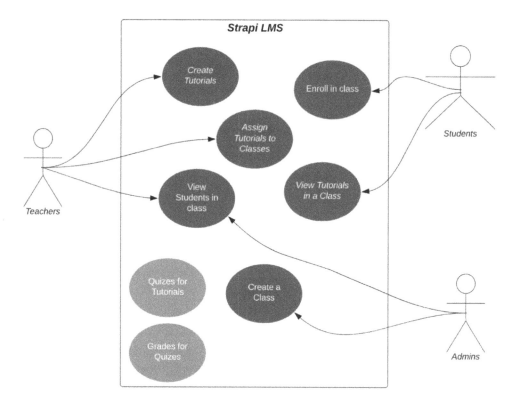

Figure 7.12: Use case diagram for main actors and functionalities of the system

As illustrated in the diagram, each user (role) should have certain permissions to interact with the API entities. For example, **Teachers** can create tutorials and edit their own tutorials but not others, and they cannot create classrooms. **Students** can view a tutorial but cannot create one, while **Admins** can perform all **CRUD** (short for **Create, Read, Update, and Delete**) operations. The following table puts all those permissions into perspective with the tutorial and classroom content-types:

Actor (Role)	Classroom Permissions	Tutorial Permissions
Students	find findOne findTutorial	findOne
Teachers	find findOne findTutorial	findOne create
Admins	All	All

Based on this table, it is clear that we need to define three roles in our API. Let's do that now.

Creating the Student role

The first role we will create is the Student role with simple permissions to view classroom and tutorial content-type:

1. From the Strapi admin panel, click **Settings** and then **Roles** under **USERS & PERMISSIONS PLUGIN**.

2. Click the **Add new role** button to create a new role.

3. Enter Student for the **Name** field and any description in the **Description** field.

4. In the **Permissions** section, tick **find**, **findOne**, and **findTutorials** for the **Classroom** content-type and **find** and **findOne** for the **Tutorial** content-type.

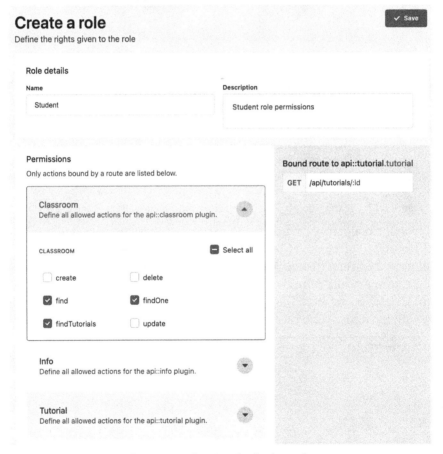

Figure 7.13: Creating the Student role

5. Click the **Save** button to save the changes.

We should have a new Student role added to the system. Next, let's create the Teacher role.

Creating the Teacher role

The Teacher role should have similar permissions to the Student role but with one extra permission for the `create` tutorial:

1. From the Strapi admin panel, click **Settings** and then **Roles** under **USERS & PERMISSIONS PLUGIN**.

2. Click the **Add new role** button to create a new role.

3. Enter `Teacher` for the **Name** field and any description in the **Description** field.

4. In the **Permissions** section, tick **find**, **findOne**, and **findTutorials** for the **Classroom** content-type and **findOne** and **create** for the **Tutorial** content-type.

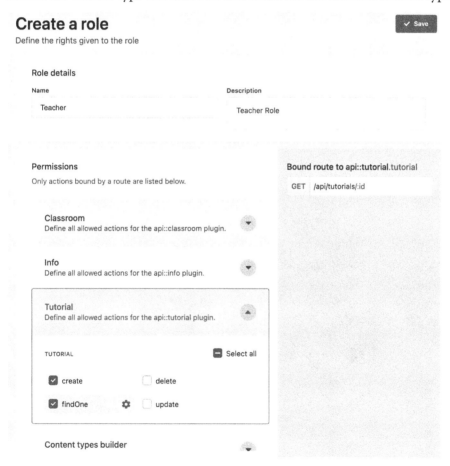

Figure 7.14: Creating the Teacher role

5. Click the **Save** button to save the changes.

We should have a new **Teacher** role added to the system. The final role we need to add is the **Admin** role. Let's do that next.

Creating the Admin role

Unlike the previous two roles we created, we will grant the **Admin** role access to everything:

1. From the Strapi admin panel, click **Settings** and then **Roles** under **USERS & PERMISSIONS PLUGIN.**

2. Click the **Add new role** button to create a new role.

3. Enter `Admin` for the **Name** field and any description in the **Description** field.

4. In the **Permissions** section, tick all permissions for both the **Classroom** content-type and the **Tutorial** content-type.

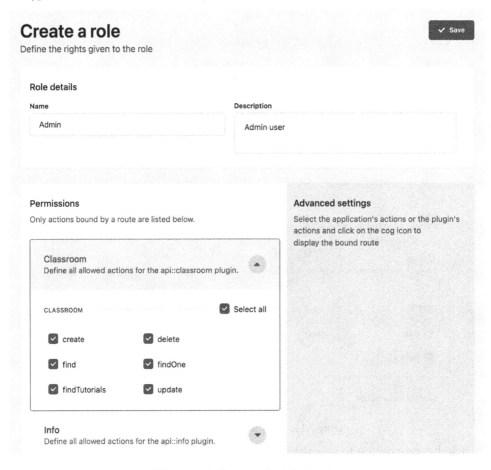

Figure 7.15: Creating the Admin role

5. Click the **Save** button to save the changes.

At this stage, we should have three new roles added to our API. The **Roles** main screen should display **0 users** next to the new roles we created.

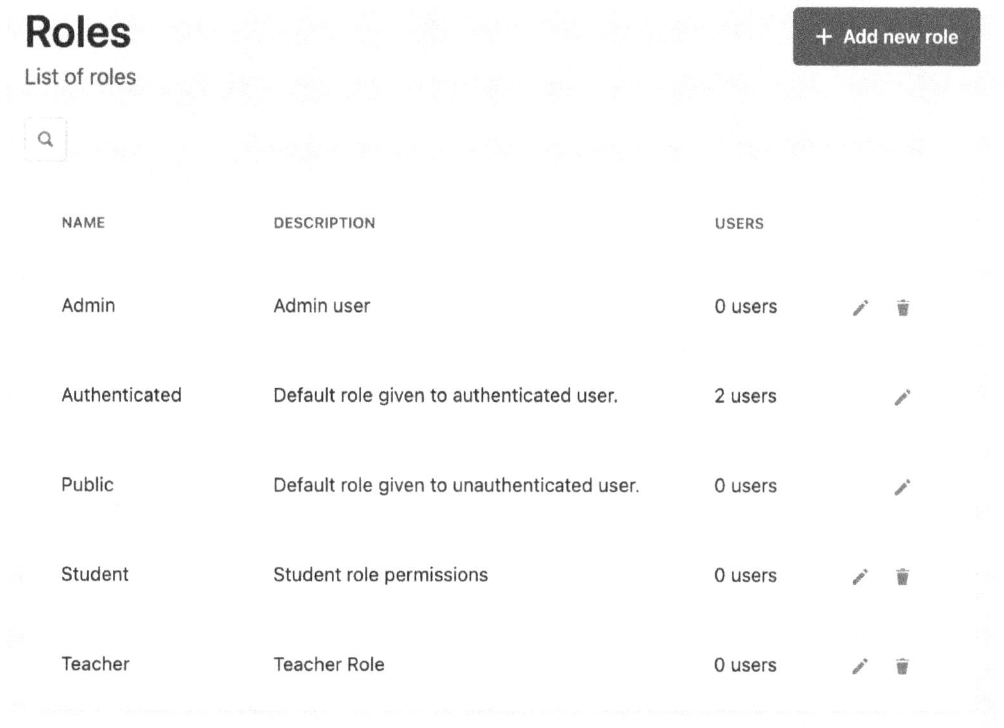

Roles

List of roles

+ Add new role

NAME	DESCRIPTION	USERS		
Admin	Admin user	0 users	✏	🗑
Authenticated	Default role given to authenticated user.	2 users	✏	
Public	Default role given to unauthenticated user.	0 users	✏	
Student	Student role permissions	0 users	✏	🗑
Teacher	Teacher Role	0 users	✏	🗑

Figure 7.16: All roles in the API

If you recall earlier in this chapter, we created an **educator1** user. This user was assigned the **Authenticated** role as this is the default behavior for Strapi – which we will fix soon – when creating a new user. For now, let's manually change that user role from **Authenticated** to **Teacher**:

1. Click on **Users** under **Collection Types**.
2. Select the **educator1** user from the users list.

3. In the **Role** drop-down menu, change the role to **Teacher**.

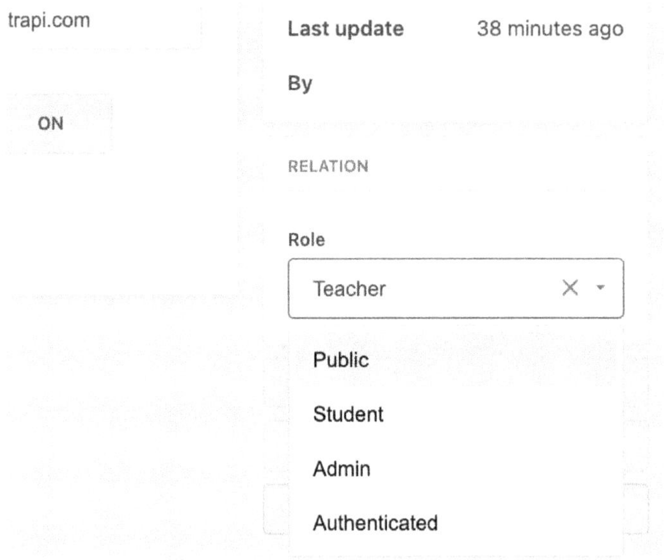

Figure 7.17: Changing the user role

4. Click the **Save** button to save the changes.

To confirm that our new permission system is working as expected, try logging in with the **educator1** user and creating a new classroom by sending a request to POST /api/ classrooms. Since the **Teacher** role is not authorized to create a classroom, we should see a **Forbidden 403** response.

What if I want to change the default role for a new user?

In the previous example, we had to change the user role manually since the default behavior for Strapi is to set the Authenticated role as the default user role. Let's change this behavior to set the default role to Student instead of Authenticated:

1. Click on the **Settings** menu.
2. Click on **Advanced Settings** under **USERS & PERMISSIONS PLUGIN**.
3. Under **Default role for authenticated users**, select **Student** from the drop-down menu.
4. Click the **Save** button to save the changes.

Now let's test this change by creating a new user account. Similar to what we did before with the `educator1` user, let's use Postman and send a `POST` request to `/auth/local/register` with the following as the JSON payload:

```
{
  "username": "student1",
  "email": "student1@strapi.com",
  "password": "password"
}
```

Notice the user role in the response. It should be set as **Student**.

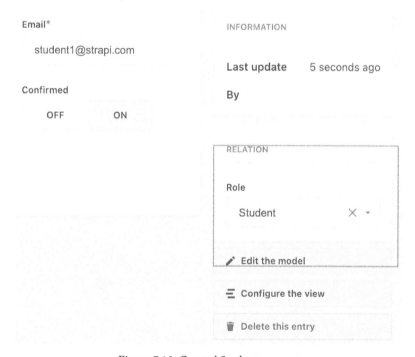

Figure 7.18: Created Student account

We have successfully implemented role-based authentication in our API with simple tweaks to the roles and permissions in Strapi. However, you might have noticed there is something missing with the Teacher role. A Teacher can create a tutorial but they cannot edit or delete it. If we granted a teacher edit and delete permissions, then basically they will be like an admin and they can edit or delete any tutorial. We do not want this, of course, and we want a teacher to be able to modify their own tutorial only.

This problem can be solved by the use of policies in Strapi. Let's discuss Strapi policies in the next section to understand what they are and how can we use them to assist us with improving the authorization flow in our API.

Working with policies

A common use case in API development is to allow users to manage their own information and contents. While the **Users & Permissions** plugin works great for providing us with role-based authentication, it does not provide an out-of-the-box strategy for allowing users to manage their own content. Instead, the decision is left to developers to handle this case according to the business requirements. Luckily, we do not need to reinvent the wheel to implement such logic in the API authentication flow. We can use Strapi's policies to customize the authentication and authorization flow.

A policy in Strapi is a function that can be executed before a request reaches a controller. Policies are mostly used for securing business logic easily. Policies are applied to a route using the `policies` array in the `{content-type}/routes/{content-type}.js` router file or the `{content-type}/routes/{custom-file}.js` custom router file:

```
// Example of policy in Core Router file
module.exports = createCoreRouter('api::tutorial.tutorial, {
  config: {
   find: {
    policies: ['is-manager', 'global::is-admin'],
    },
   },
});


// Example of policy in Custom Router file
{
  "method": "GET",
  "path": "/classrooms/:id/tutorials",
  "handler": "classroom.findTutorials",
  "config": {
   "policies": ["is-manager", "global::is-admin"]
  }
},
```

There are two types of policies in Strapi: a global policy or an API policy. A global policy, as the name suggests, is intended to be used with multiple API routes. For example, a policy that checks whether a user is logged in can be used by both the classrooms routes and the tutorials routes. Global policies are created in the `/src/policies` folder. In the previous code snippet, `is-admin` is an example of a global policy. Notice we used the `global::` prefix before the policy name.

An API policy is a policy that is intended to be used within specific API routes. For example, we might have a policy that checks whether a user is allowed to edit a classroom. Such a policy will only be used within the classroom API routes; therefore, it can be created as an API policy. The API policies are stored in the `/src/api/[api-name]/policies` folder. In the previous code snippet, `is-manager` is an example of an API policy.

Creating a policy

Going back to our problem of allowing editing and deleting, let's assume we have the following requirements:

As a classroom manager

I want to be able to edit or delete tutorials I created for my classroom

So that I can manage my classroom effectively

To solve this problem, we are going to create a policy called `is-manager` that will allow a user to delete or edit tutorials if they are the classroom manager.

There are two ways to create a policy in Strapi. The first method is to use the Strapi CLI and the `generate:policy` command. The second method is by manually creating a policy file, so for this example, we would create an `is-manager.js` file in the `policies` folder. Note that the policy name must match the filename. We are going to use the CLI to generate the policy.

The policy is going to be used within the tutorials routes only, so we are going to create an API policy. Let's get started:

1. From the command line, issue the `yarn strapi generate policy` command and follow the onscreen prompt to generate the policy.

2. Enter `is-manager` for the policy name.

3. Select **Add policy to an existing API** when asked where you want to add the policy to.

4.   Select the **tutorial** API.

```
~/code/nyala/Building-APIs-with-Strapi  ⌥ chapter_7 ±   yarn strapi generate policy
yarn run v1.22.17
$ strapi generate policy
? Policy name is-manager
? Where do you want to add this policy? Add policy to an existing API
? Which API is this for? tutorial
✔  ++ /api/tutorial/policies/is-manager.js
✦  Done in 7.82s.
```

Figure 7.19: Generating the is-manager policy

Once you are done, a new `src/api/tutorials/policies/is-manager.js` file will be generated. Let's examine this file:

```
'use strict';
/**
 * `is-manager` policy.
 */
module.exports = (policyCtx, config, { strapi }) => {
    // Add your own logic here.
    strapi.log.info('In is-manager policy.');
    const canDoSomething = true;
    if (canDoSomething) {
        return true;
    }
    return false;
};
```

The file exports a default function that accepts three parameters. The first parameter, `policyCtx`, is the most interesting parameter to us. This is a wrapper around the controller context.

> **Remember**
>
> Every controller action receives a context object as a parameter. This object contains the request context and the response context. It also holds the state object where useful information, such as the current user, is kept.

The function body itself contains a simple log and a dummy implementation to help us get started. Generally speaking, a policy should return either `true` or `false`. If the policy function returns `true`, then the request will be passed to the next policy, if any, or reach the controller action. If the policy function returns `false`, then the request will not reach the controller action and will be rejected.

5. This policy is not used with any controller actions, so let's update the tutorial router file to apply the policy. Edit the `/src/api/tutorials/routes/tutorial.js` file by adding the `is-manager` policy, as shown:

```
You, a minute ago | 1 author (You)
'use strict';

/**       You, a week ago • Chapter 3
 * tutorial router.
 */

const { createCoreRouter } = require('@strapi/strapi').factories;

module.exports = createCoreRouter('api::tutorial.tutorial', {
  except: ['find'],
  config: {
    update: {
      policies: ['is-manager'],
    },
    delete: {
      policies: ['is-manager'],
    },
  },
});
```

Figure 7.20: Applying a policy to a route

6. Before we test it out, make sure to update the permissions for the Teacher role, allowing `update` and `delete` on the `/api/tutorials/:id` endpoint.

7. We do not want to delete anything yet; we just want to test that the policy is configured properly. Let's use a random ID that does not exist at the tutorials DELETE endpoint. In Postman, issue a DELETE request to the `/api/tutorials/:id` endpoint: DELETE `/api/tutorials/1000`.

8. Check the Strapi console logs. You should see the `In is-manager policy` message printed before the request debug message.

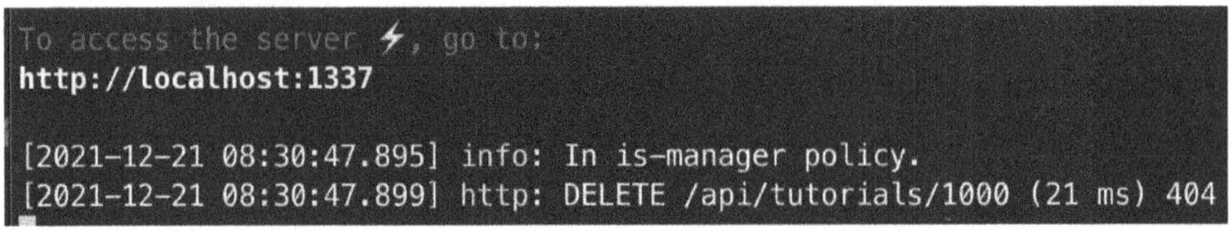

```
To access the server ⚡, go to:
http://localhost:1337

[2021-12-21 08:30:47.895] info: In is-manager policy.
[2021-12-21 08:30:47.899] http: DELETE /api/tutorials/1000 (21 ms) 404
```

Figure 7.21: Policy debug message

> **Note**
>
> The API should return 404 as the 1000 tutorial ID does not exist in the system. The purpose here was to ensure the policy is configured correctly and to demonstrate that it will be executed before the request.

9. Next, let's edit the `is-manager` file, adding our business logic:

```
"use strict";


/**
 * `is-manager` policy.
 */

module.exports = async (policyCtx, config, { strapi }) =>
{
  try {
    // Get id from request
    const { id } = policyCtx.params;


    // Get the tutorial by id
    const tutorial = await strapi.service('api::tutorial.
    tutorial').findOne(id, { populate: ['classroom.
    manager']});


    // current logged in user
    const { user } = policyCtx.state;


    // check if classroom manager is logged in user
    const { classroom: { manager } } = tutorial;
```

```
    if (manager&& manager.id === user.id) return true;

    return false
  } catch (e) {
    strapi.log.error("Error in is-manager policy", e);
    return false;
  }
};
```

The policy works by getting the tutorial from the database and the currently logged-in user from the context state object. If the current user is the classroom manager, the request will be passed to the controller; otherwise, the request will be denied and an unauthorized error will be returned to the user.

10. Make sure you are logged in with the **educator1** user we created at the beginning of this chapter and let's try deleting the first tutorial we created in the previous chapters. From Postman, send a request to delete the tutorial with an ID of 1.

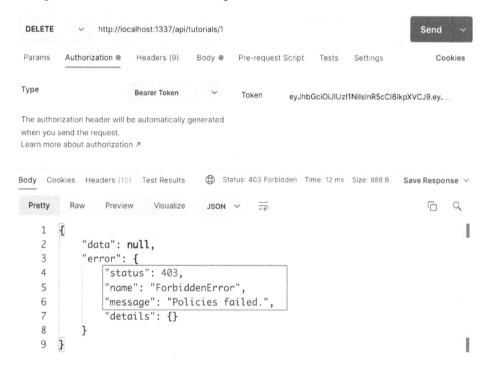

Figure 7.22: Unauthorized action error message

Now, let's change the classroom manager to allow deleting the contents:

1. From the Strapi admin panel, find the **Strapi 101** classroom and change the manager from **testuser** to **educator1**.

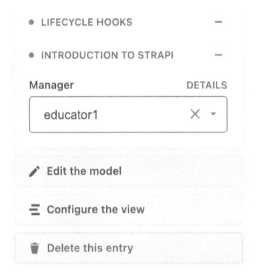

Figure 7.23: Changing classroom manager

2. Save the changes, wait for the server to restart, and try deleting the same tutorial again.

3. This time, you will be allowed to delete the tutorial.

Policies in Strapi are a great mechanism to enhance the authorization flow. They can be applied to single or multiple routes, which helps in enforcing the **DRY** (short for **Do Not Repeat Yourself**) principle.

Honorable mention – OAuth providers

As we have seen so far in this chapter, it is quite easy to set up an authentication system in Strapi just by using the **Users & Permissions** plugin. It is also worth mentioning that Strapi allows OAuth providers to integrate authentication in the API easily.

The configurations for the OAuth providers can be found on the **Settings** page under the **USERS & PERMISSIONS PLUGIN** section.

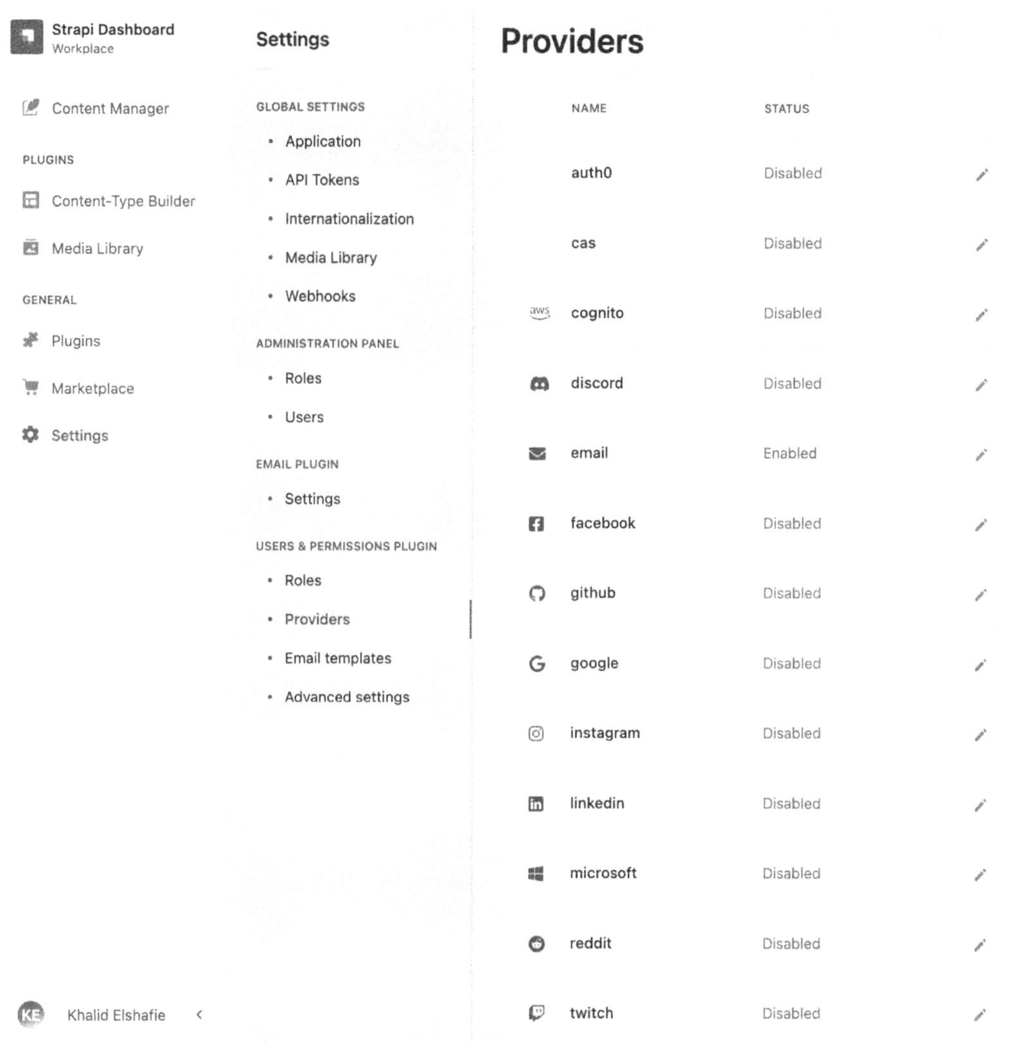

Figure 7.24: OAuth providers list

The configuration is straightforward; in most cases, you will need to obtain a client ID and a client secret from the provider you want to use to apply it in the admin panel. The Strapi documentation provides an example for each of the providers and how to obtain the required secrets. We feel there is nothing more we can add to it.

Summary

We started this chapter by explaining the difference between Strapi admin users and API users. Afterward, we explored how API users can sign up and log in to the API using the routes exposed by the **users-permissions** plugins. We also discussed JWTs and saw how we can edit their configurations.

After that, we started working on securing and protecting the API routes. We changed the `create` tutorial's endpoint, making it accessible by logged-in users only, and explained how the authorization flow works in Strapi.

Then, we took a deep dive into the roles and permissions. We revisited the main actors diagram for our API and used it to create the required roles by our API.

Finally, we introduced the concept of policies in Strapi and saw how can we use them to further enhance the API authorization flow by allowing users to edit and delete their own content only.

In the next chapter, we will explore the Strapi plugin system. We will understand the Strapi plugin ecosystem and how we can enable, disable, and configure plugins.

8
Using and Building Plugins

In this chapter, we will explore Strapi plugins. Plugins open unlimited possibilities for Strapi—they allow us to expand Strapi with new capabilities and functionality in a way that is reusable to avoid reinventing the wheel, and they also allow us to tap into the experiences of the larger Strapi open-source community. This functionality ranges from the basics of maintaining content, augmenting it with comments, to advanced use cases such as migrating data or syncing roles and permissions between environments. We will solve two common requirements that can be tackled using plugins. The first is enabling GraphQL for our **API** (short for **application programming interface**), which is becoming an increasingly popular alternative to **REpresentational State Transfer** (**REST**) APIs. The second is sending emails from Strapi.

Here are the topics that we will cover in this chapter:

- Strapi plugin ecosystem
- Installing and using plugins from the **Marketplace**
- Use case—Enabling GraphQL for our API
- Use case—Sending an email from Strapi
- Creating our own plugins

Exploring the Strapi plugin ecosystem

Strapi aims to be easy to use and flexible to extend. As we saw repeatedly in the previous chapters, it's very easy to build an API for the default **create, read, update, and delete (CRUD)** operations, but we are not locked into the default behavior—it's always possible to extend the default behavior for our API in any way we want. That philosophy goes beyond building and coding APIs and their interactions, to almost every aspect of Strapi and its admin panel.

To achieve such flexibility, Strapi builds heavily on a plugin's architecture to open possibilities for infinite use cases beyond the core functionality.

What is a plugin in Strapi?

In the admin panel, if we navigate to **Plugins** under **General** in the main menu, we will see that there are a few plugins already installed. We have already interacted with most of these, and more detail on these is provided here:

- **Content Manager**: This is where we edit our classroom, tutorial, and other entities.

- **Content Type Builder**: This is where we define our content-types.

- **Roles & Permissions**: This is where we define roles and permissions associated with our API endpoints.

- **Media Library**: This is where we attach media to our content.

- **Internationalization**: This is where we are able to define different locales for our content.

- **Email**: This is the only plugin in the list we haven't used, but we are going to do so in this chapter.

- The aforementioned plugins are shown in the following screenshot:

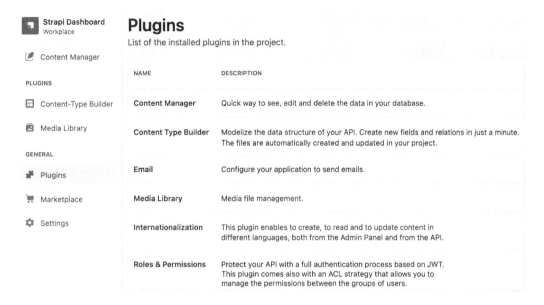

Figure 8.1: List of installed plugins

It might be a bit paradoxical that the **Content Manager** and **Content Type Builder** are in the plugins list. After all, they are not optional—they are at the core of what Strapi (or any **content management system** (**CMS**)) does. What's left in a CMS if it can't manage content?

From a system design and architecture point of view, though, it is important to understand—and appreciate—the decision of Strapi's creators to have this functionality developed as a plugin. This opens the door for the open-source community and everyone else to extend Strapi and for their extensions to be treated the same way as other plugins (including default core ones) by Strapi.

Before getting into the details of how plugins work in Strapi, let's explore some of the plugins that are provided by Strapi and learn how to install and use them.

Installing and using plugins from the Marketplace

Strapi has a **Marketplace** that curates some of the official Strapi plugins to install directly from the admin panel. At the time of writing this book, the **Marketplace** is undergoing major changes and upgrades. The following screenshot shows version 3 of the Strapi **Marketplace**:

Figure 8.2: Strapi Marketplace version 3

The **Marketplace** will host a couple of official Strapi plugins that are not installed by default, as well as community-developed plugins. One of the official Strapi plugins that are not installed by default when creating a new Strapi project is the **API Documentation** plugin. This plugin allows us to create an **OpenAPI** document and visualize our API with the **Swagger** UI. OpenAPI is a standard for describing **REST** APIs that helps when documenting APIs, visualizing them with tools such as **Swagger** UI, automating them, and more.

While we can directly install a plugin from the **Marketplace** by clicking the **Download** button, since the **Marketplace** is not available yet at the time of writing this book, we are going instead to use the Strapi **command-line interface** (**CLI**) to install the documentation plugin.

Installing the API Documentation plugin from the CLI

Follow these steps to install the API **Documentation** plugin:

1. To install the **Documentation** plugin, all we need is to run the following command at the root of our project:

    ```
    yarn strapi install documentation
    ```

    You can see the output of the command in the following screenshot:

    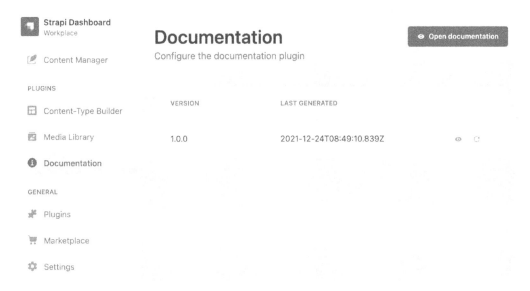

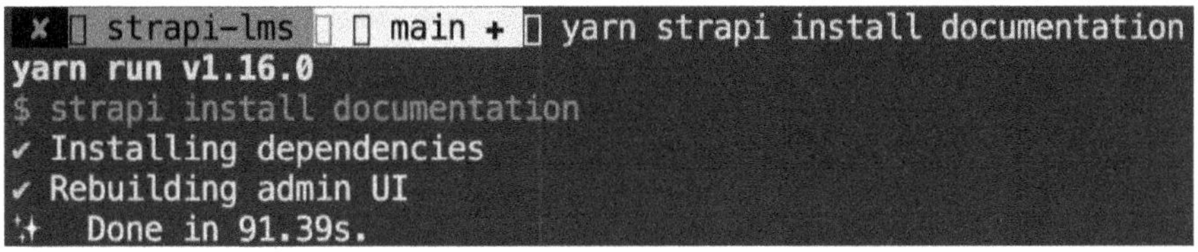

    Figure 8.3: We can install plugins from the CLI

2. Once the installation is complete and our server restarted, we will get a new entry for our **Documentation** plugin in the menu under **Plugins**. Click on the **Documentation** plugin. This will open the settings of our plugin and will show an **Open documentation** button, as illustrated in the following screenshot:

Figure 8.4: The settings page for the Documentation plugin we installed

> **Note**
>
> The **Documentation** menu item should appear in the main menu once the installation is complete and the server restarts. However, if you do not see the **Documentation** menu item, try restarting the server manually.

3. By clicking the **Open documentation** button on the top right of the page, we will be taken to **Swagger** UI where we can read the documentation for our API and interact with it, as illustrated in the following screenshot:

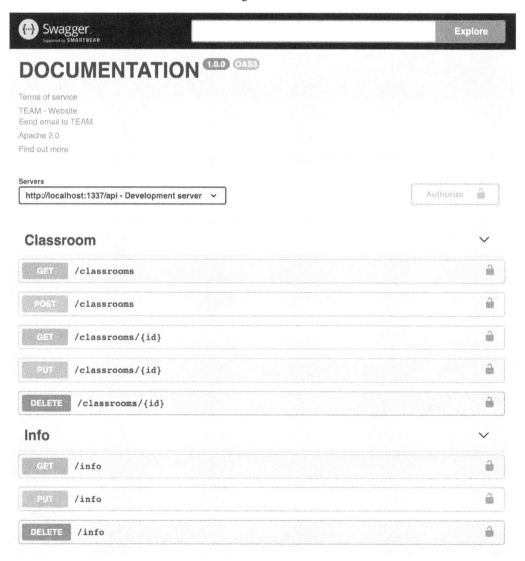

Figure 8.5: Using the Documentation plugin, we can visualize our API using Swagger UI

4. If we would like to delete the plugin, then we can run the following command in the CLI:

```
yarn strapi uninstall documentation --delete-files
```

> **Cleaning Up Generated Documentation**
>
> Note that this plugin generates the documentation in the filesystem, which is great as we can add it to our source control and share it with our team. When we uninstall the plugin, these files are still there, though. If we want to get rid of them, then we will have to do this manually. The files will be in a `documentation` folder for all the APIs we have—for example, the documentation for our classroom endpoint will be under `src/api/classroom/documentation/1.0.0/classroom.json`.

We'll see how to find more plugins on the **Marketplace** in the next section.

Finding more plugins

Marketplace is likely to mature and grow to become the de facto place for finding and installing plugins once its development is complete. The documentation page listing the available plugins is also still a work in progress. There are a couple of other places where we can find and explore more plugins, though.

Awesome Strapi is a semi-official list of curated plugins on GitHub: `https://github.com/strapi/awesome-strapi`. The list is not limited only to plugins, but also includes links to tutorials and other useful resources.

Another place to look for plugins is the **Node Package Manager** (**npm**) package registry. Strapi plugins are named following a certain format. They have to be prefixed with `strapi-plugin-*` (that is, `strapi-plugin-documentation`), so if we want to browse all Strapi plugins, we can search for packages starting with that prefix or go directly to `https://www.npmjs.com/search?q=strapi-plugin`, and we will see a list of all the plugins developed for Strapi.

As with any open-source project, some of the plugins might be unmaintained or outdated, so make sure to check the repository's stars, activity, and documentation to get a good indication of the quality of plugins before adding them to your project.

Now that we have explored how to find and install plugins, let's spend some time exploring two popular plugins: the **GraphQL** and **Email** plugins.

Use case – Enabling GraphQL for our API

So far, we've been consuming our API through a RESTful interface. We've added support for generating documentation in the previous section through the **Documentation** plugin. GraphQL is becoming an increasingly popular alternative to REST, and one of its many benefits is that generates schemas for our types that serve as documentation that is always up to date and reflects the current state of our content-types.

To support GraphQL in our Strapi API, we can simply install a plugin that enables a GraphQL endpoint to interact with our content and automatically generates all the GraphQL schema for our content-types.

Let's go ahead and install the plugin with the following command:

```
yarn strapi install graphql
```

Now, if we start our Strapi instance with `yarn develop` and open our browser at `http://localhost:1337/graphql`, we should see GraphQL Playground in the browser. GraphQL Playground is a visual interface that allows us to write GraphQL queries and mutations to interact with our content. You can see an illustration of this interface in the following screenshot:

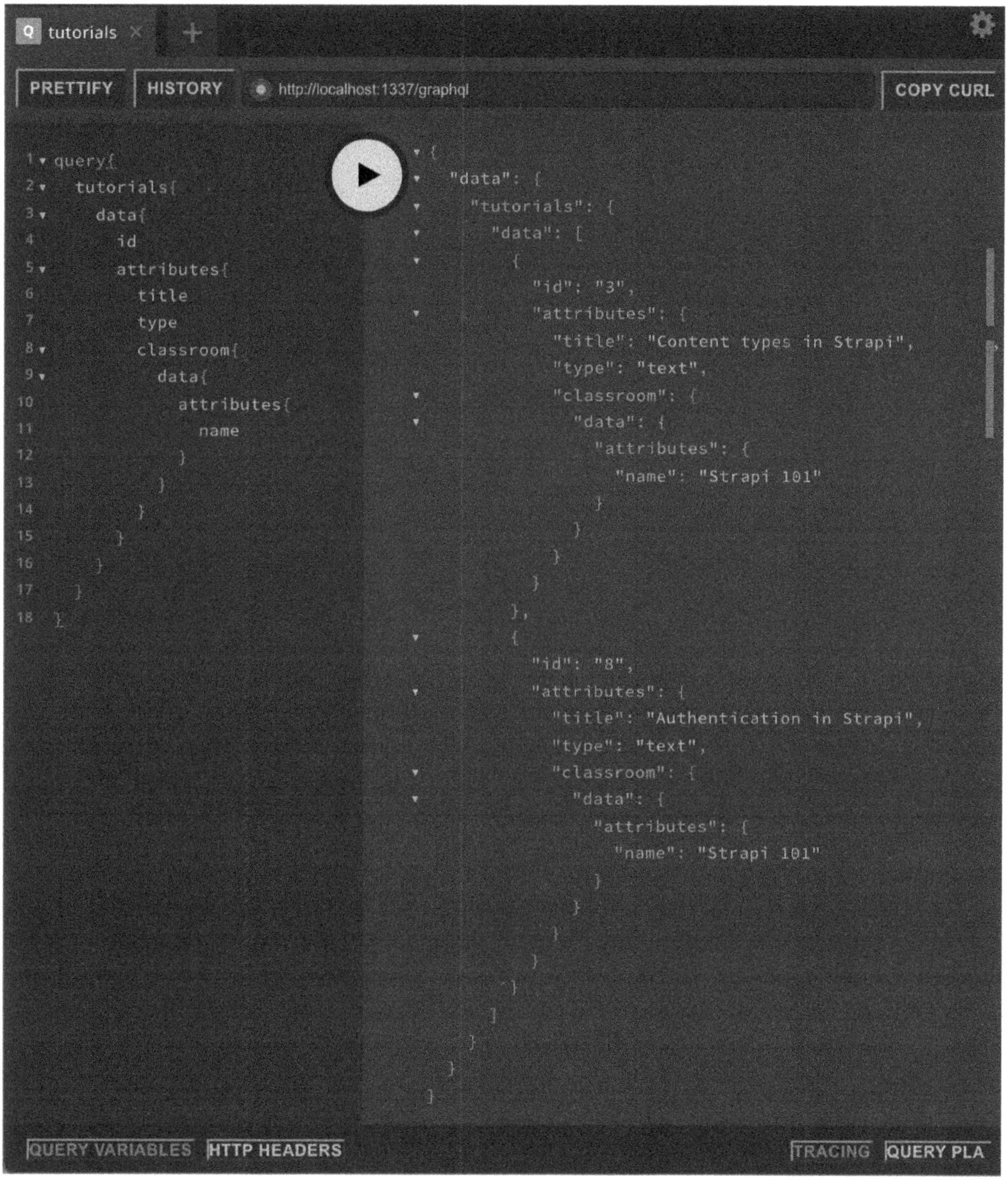

Figure 8.6: The GraphQL Playground

In GraphQL Playground, we can—for example—query all tutorials with their classrooms and the managers of these classrooms by sending a GraphQL query such as this:

```
query{
  tutorials{
```

```
    data{
      id
      attributes{
        title
        type
        classroom{
          data{
            attributes{
              name
            }
          }
        }
      }
    }
  }
```

Using the REST API, this would require sending a second query or building a custom route to retrieve the manager information, as Strapi defaults to generating one level of relationships by default (that is, it will generate the classroom related to the tutorials but not the manager related to the classroom). With GraphQL, we have total control over which fields we want to query and which entities to populate.

For each model, the plugin autogenerates queries and mutations that are ready to use on our content-types. We can search the documentation from GraphQL Playground to find all possible interactions and the expected schema—there is no need for the **Documentation** plugin, with the risk of the documentation going stale. For example, if we click on **Docs** on GraphQL Playground and search for createTutorial, we can see the documentation for the mutation, as illustrated in the following screenshot:

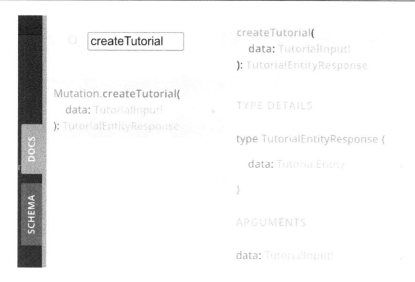

Figure 8.7: Exploring the documentation from GraphQL Playground

Now, let's try to create a tutorial from the GraphQL endpoint sending the mutation, as follows:

```
mutation createTutorial {
  createTutorial (data: { title: "GraphQL in Strapi",
  slug:"graphql-in-strapi", type:"text", classroom: 1 }) {
    data {
      id
      attributes {
        title
      }
    }
  }
}
```

We will get a 404 Forbidden response because this is an unauthenticated request.

To authenticate with GraphQL, we can send a mutation passing our **identifier** (**ID**) and password, as follows:

```
mutation login{
  login(input: { identifier: "educator1@strapi.com",
  password: "password" }) {
    jwt
```

```
        }
    }
```

> **Note**
>
> Make sure to use valid user information and that the user has the correct permissions to create a tutorial. Refer to *Chapter 7, Authentication and Authorization in Strapi*, for more details.

This will return a **JavaScript Object Notation (JSON) Web Token (JWT)** token. Then, we can use the token as part of an Authorization header (in the **HyperText Transfer Protocol (HTTP)** headers section of GraphQL Playground). The header will look like this: `{ "Authorization": "Bearer JWT_TOKEN" }`. Here, `JWT_TOKEN` is the token returned by the login mutation. Once we add the header, we can rerun the `createTutorial` mutation, and the classroom will be created successfully. The following screenshot illustrates the process:

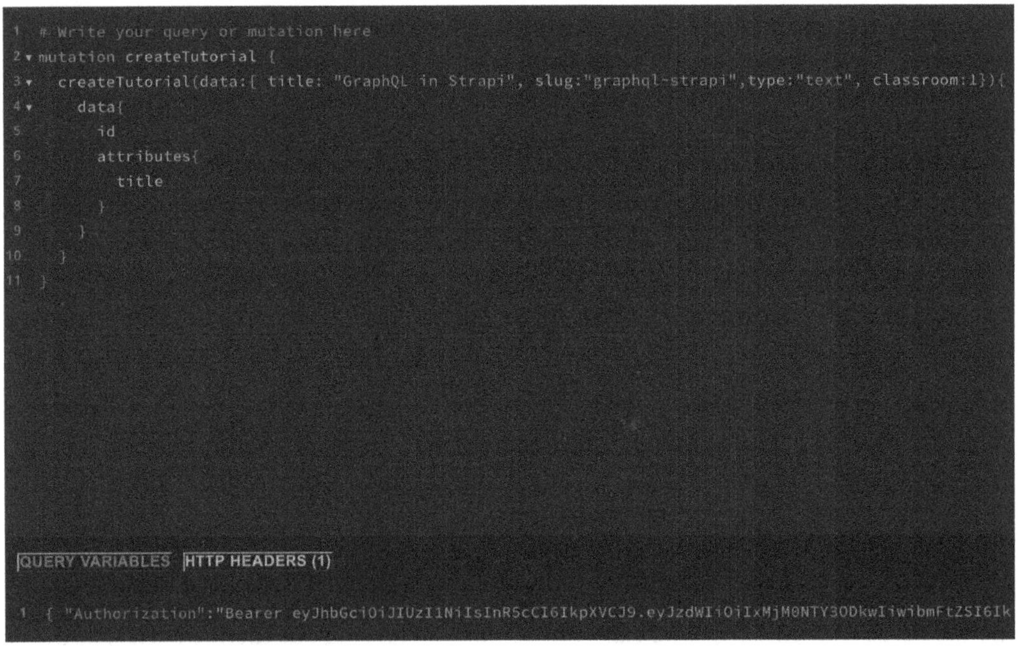

Figure 8.8: Adding an Authorization header to our GraphQL mutation

The autogenerated schema is likely to cover most of our use cases, especially given the inherent flexibility of GraphQL. As with everything in Strapi, though, if we want to change the behavior, then we are not stuck. The **GraphQL** plugin provides the ability to customize the schema by defining new types, custom resolvers, disabling certain mutations or queries, and more. Strapi's documentation for the **GraphQL** plugin is excellent in case you need to extend the default behavior.

Now that we have explored using the **GraphQL** plugin, let's move on to another popular use case that we can solve with a plugin: sending emails from Strapi.

Use case – Sending an email from Strapi

One plugin installed by default that we haven't used so far is the **Email** plugin. This is a plugin that allows us to send emails from Strapi. It exposes a service that can be used from our API. Let's imagine a scenario where we have a requirement such as this:

As a class admin,

I want to be notified by email when a new tutorial is created

So that I can keep track of the activities in our system.

Before hooking our email notification to tutorial creation, let's first see how we can send an email from Strapi.

Sending an email using the Email plugin

The **Email** plugin provides a service that we can use to send emails. We created our own service before in *Chapter 5*, *Customizing Our API*, to encapsulate logic for generating a summary. Similarly, plugins can expose services that allow us to use their functionality from other layers in the API. Plugins, as with API endpoints, make use of the same building blocks: controllers, routes, models, and all other concepts we've used so far.

To send an email, we can write a code snippet similar to this:

```
await strapi.plugins['email'].services.email.send({
  to: 'mozafar@nyala.dev',
  from: 'khalid@nyala.dev',
  subject: 'Sending an email with Strapi ',
  text: 'Hello!',
  html: '<h1>Hello!</h1>',
});
```

To explore the settings of the plugin, we can navigate to **Email Settings** under **Settings**. The page also includes a direct link to the documentation of the plugin, where we can see a full list of options we can pass to the send method, as illustrated in the following screenshot:

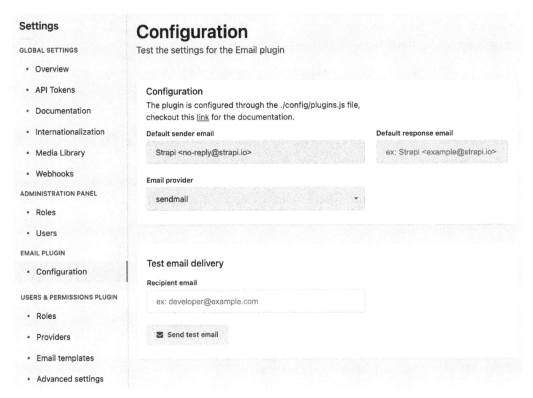

Figure 8.9: Email plugin settings

One setting is the email provider, which is set to sendmail right now. The **Email** plugin uses a sendmail local email system by default. We can instruct it to use a different provider, as we will see in the next section.

Plugins and providers

Related to the concept of plugins in Strapi is the concept of providers. Providers are further configuration points for plugins.

Some examples of providers we encountered before are the **Open Authorization** (**OAuth**) providers used by the **Users & Permissions** plugin in the previous chapter. The **Upload** plugin (which is used internally to upload images and assets in the Media Library), can also be configured to use a different provider to upload assets to **Amazon Simple Storage Service** (**Amazon S3**) or another storage service.

For the **Email** plugin, we can configure it to use providers for external email services such as **SendGrid** or **Amazon Simple Email Service** (**Amazon SES**).

To find providers available for the **Email** plugin, we can check the documentation or search for npm packages starting with `strapi-provider-email-*`. By default, providers will follow the `strapi-provider-[pluginName]` naming convention. You can see a list of available email providers here:

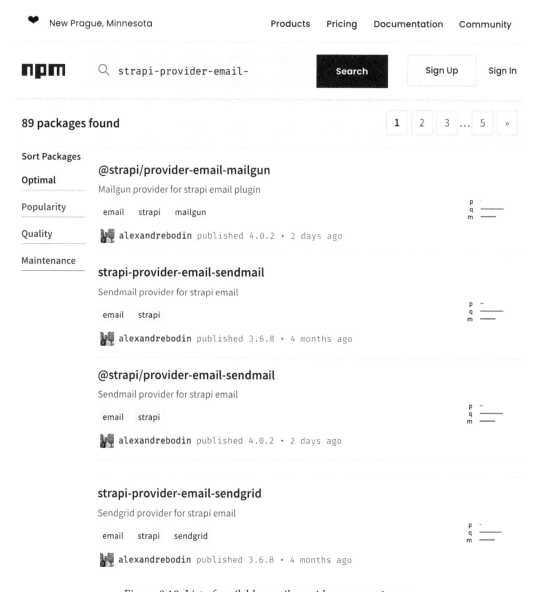

Figure 8.10: List of available email providers on npmjs.com

As we can see, there are packages for a variety of email providers. To explore how we can set one up, let's install the provider to enable the **Mailgun** service. Installing and configuring any provider will likely consist of the same steps, as outlined here:

1. Install the package for the email provider.

2. Configure the provider in the **config** section for the plugin.

    So, let's install the **Mailgun** plugin and configure it for our API.

3. We install the **Mailgun** plugin with the following command:

    ```
    yarn add @strapi/provider-email-mailgun
    ```

4. Then, we can update the configuration of the plugin to use our newly installed provider. To do so, create a new config/plugins.js file, but in our case, let's make the configuration specific to one environment (similar to what we learned in *Chapter 2, Building Our First API*) by adding it to config/env/production/plugins.js, as follows:

    ```
    // ./config/env/production/plugins.js
    module.exports = ({ env }) => ({
        email: {
          config: {
           provider: 'mailgun',
           providerOptions: {
              apiKey: env('MAILGUN_API_KEY'),
              domain: env('MAILGUN_DOMAIN'),
             host: env('MAILGUN_HOST', 'api.mailgun.net'),
             //Optional. If domain region is Europe use
               'api.eu.mailgun.net'
           },
           settings: {
              defaultFrom: 'myemail@protonmail.com',
              defaultReplyTo: 'myemail@protonmail.com',
           },
          },
        },
    });
    ```

By adding this config file under `config/env/production`, we instructed our API to use `mailgun` in production and use the default `sendmail` in development mode. The provider settings themselves are available in the documentation for the `@strapi/provider-email-mailgun` package.

5. To test our new provider (provided you have created an account with `mailgun`, which is out of the scope of our book), we will have to run our API in production mode with the following command:

    `NODE_ENV=production yarn develop`

    We should get an error from the `mailgun` provider that we did not provide an API key and domain. The code we added to `plugins.js` refers to `env("MAILGUN_API_KEY")` and `env("MAILGUN_DOMAIN")`, which are the environment variables that we need to provide.

6. To create these environment variables, use the `.env` file in the root of the project (which is in the `.gitignore` list by default so that our secrets are not exposed). The code is illustrated in the following snippet:

    ```
    # .env file
    MAILGUN_API_KEY=xxxx # The key from your mailgun account
    MAILGUN_DOMAIN=xxxx # The domain value from your mailgun
    account
    ```

Now, if we rerun our API in production mode, then it should run without errors, and we are ready to use our new provider to send an email. Now, let's go back to our original requirement to send an email when a new tutorial is created.

Hooking our API to send an email

The route for creating a tutorial is `POST /tutorials`. There are several points where we could extend the logic for this route to send an email. We can implement a custom controller for the route, but we also want to make sure that an email is sent whether the tutorial is created from the API or the admin panel. To achieve that, we can make use of the model `lifecycle` methods and send an email *after* the tutorial is created in the database in the `afterCreate` hook.

We can extend the tutorial model like this:

```
// api/tutorial/models/tutorial.js
"use strict";
```

```
/**
* Read the documentation (https://strapi.io/documentation/
developer-docs/latest/development/backend-customization.
html#lifecycle-hooks)
* to customize this model
*/

module.exports = {
  lifecycles: {
    //    ... other lifecycle methods
    async afterCreate(event) {
      try {
        const { result } = event;
        const classManager = result.classroom &&
        result.classroom.manager;

        if (!classManager || !classManager.email) {
          strapi.log.info(
            "No manager set for this classroom, so not
            sending an email."
          );
          return;
        }

        const { firstname } = result.createdBy;
        const subject = `A new tutorial was created by
        ${firstname}`;
        const text = `A new tutorial titled
        "${result.title}" was created by ${firstname}`;

        const email = {
          to: classManager.email,
          subject,
          text,
        };

        strapi.log.info(
```

```
        "An email will be sent with the following
        options:",
        email
      );

      await strapi.plugins["email"]
      .services.email.send(email);
    } catch (err) {
      strapi.log.error("Failed to send an email on
      tutorial creation");
      strapi.log.error(err);
    }
  },
};
```

In the `afterCreate` hook here, we are extracting information from the event parameter to construct an email to send to the class manager. All of the information needed already exists in the `result` parameter of the `event` parameter passed to the `lifecycle` hook. We first check if there is a class manager and that the class manager has an email defined. Then, we define the email subject and text values based on the tutorial title and the user who created that tutorial.

Once we have the email information (the `to`, `subject`, and `text` fields), we can send the email. Sending the email itself is very simple—it's a one-liner using our **Email** plugin, as follows:

```
await strapi.plugins["email"].services.email.send(email);
```

We also wrapped our hook in a `try/catch` statement to make sure that failing to send the email doesn't throw an error to the user, as it's not a core operation. In larger production systems, we're likely to want to make this operation more robust by decoupling sending the email in a more resilient way, using a message broker or another solution.

So far, we have used readily available plugins to generate documentation, create a GraphQL endpoint, and send emails. Next, we will learn the fundamentals of building our own plugins so that we can enrich Strapi even further, and potentially even contribute back our amazing plugins to the Strapi community.

Creating our own plugins

So far, we've been installing **external plugins** that live in our node_modules folder. Strapi also provides us with a way to build **local plugins**. Functionality-wise, these are similar to external ones, except that they're part of our project and can't be readily reused in other projects.

So, let's create our first plugin, as follows:

1. If you are following along from the previous section, then stop the server to make sure that NODE_ENV is not set to production, then start the server using the yarn develop command.

2. To create a new plugin, we can use the CLI to generate its skeleton. Let's run the following command:

    ```
    yarn strapi generate plugin lms
    ```

    Running the preceding command will generate a plugin skeleton for us.

3. Next, we will need to enable the plugin by adding it to the config/plugins.js file. Create this file and add the following code to it:

    ```
    module.exports = {
      // ...
      'lms': {
        enabled: true,
        resolve: './src/plugins/lms'
      },
      // ...
    }
    ```

    > **Remember**
    >
    > In the previous section when we configured the **Email** plugin, we created a config/env/production/plugin.js config file because we wanted that plugin to be used in production only. However, for this plugin we just created, we added the config to config/plugins.js, which means it will be included in all environments.

4. Let's now restart our project, run it with --watch-admin to ensure the rebuilding of the admin panel as we extend it with our plugin. In the terminal, run the following command:

    ```
    yarn develop --watch-admin
    ```

> **What Is the --watch-admin Flag?**
>
> The `--watch-admin` flag is intended to be used when customizing
> the admin panel. If you recall, the admin panel is built using React.js. The
> `--watch-admin` flag runs a `webpack` server with hot reloading enabled
> to help while developing the admin panel. The `dev` server can be accessed
> using port `8080`.

5. Now, if we log in again to the admin panel (it will be running in a different port since
    it's in watch mode, likely `http://localhost:8080`), then we will see our plugin
    showing in a list of plugins in the menu, as illustrated in the following screenshot:

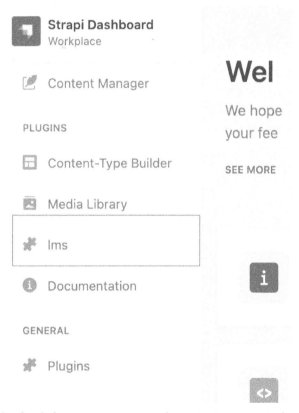

Figure 8.11: Our local plugin now appears on the navigation menu in the admin panel

6. Edit the name of the plugin from `package.json` in the `plugin` folder and change the name and description, as illustrated in the following screenshot. Those values are shown in the admin panel in the **Plugins** menu item:

```
1 ⌄ {
2      "name": "lms",
3      "version": "0.0.0",
4      "description": "A strapi plugin to help manage the LMS systen",
5 ⌄    "strapi": {
6        "name": "LMS Dashboard",
7        "description": "Dashboard for the LMS system",
8        "kind": "plugin"
9      },
10     "dependencies": {},
```

Figure 8.12: We can update the plugin metadata from the package.json strapi section

Now, after updating the `package.json` file, we will see the entry on our menu updating to the new name, and the plugin description appearing in the list of plugins, as illustrated in the following screenshot:

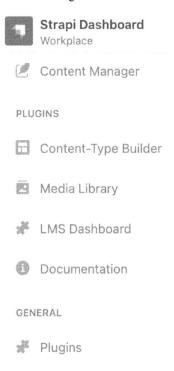

Figure 8.13: The custom name for our plugin is reflected in the admin panel menu

Before starting to build our plugin, let's have an overview of which files were generated by the CLI and briefly discuss their role.

Plugin development – frontend versus development

Looking at the files that were generated when we created our new plugin with the CLI, we can see there are some familiar folders and files. Generally speaking, a plugin development process is divided into two parts: the **backend** and the **frontend**. In the following screenshot, you can see the folder structure of the plugin code:

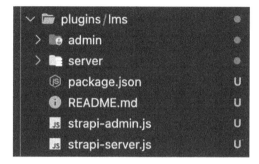

Figure 8.14: The folder structure of the plugin code

The backend components of our plugin live inside the `server` folder, which uses the same concepts we've been using with building APIs so far: routes defined under the config section, controllers to handle these routes, and services to encapsulate reusable logic. Additionally, we can create models if our plugin API needs to interact with the database. All these concepts are identical to what we have used before to build APIs. After all, the content-types that we have been using to define and create our entities are themselves part of a plugin built the same way.

The frontend components of our plugin live inside the `admin` folder and they are built using the React library.

The `admin` folder contains the entry point for our plugin (`admin/src/index.js`) that is responsible for showing our plugin in the navigation menu and registering it for use. We're unlikely to change that file much, but one common scenario to update it would be to set the permission required to view the plugin in the menu. For our scenario, the plugin will be visible to anyone with access to the admin panel.

The entry point for the actual frontend (what the user sees when they click on our plugin) lives under `admin/src/pages/App/index.js`. This module defines the frontend routes for the plugin (it uses the *React Router v5* library). Initially, it only contains one route and a fallback 404 page if we hit an unknown page. The one route points to a `HomePage` component (under `src/plugins/lms/admin/src/pages/HomePage/index.js`). Let's update this component so that it contains the following code:

```
/*
 * HomePage (plugins/lms/admin/src/pages/HomePage/index.js)
 */

import React, { memo } from "react";
import pluginPkg from "../../../../package.json";

const HomePage = () => {
  return (
    <div>
      <h1>{pluginPkg.strapi.name}</h1>
      <h2>{pluginPkg.strapi.description}</h2>
      <p>This is where our plugin will live...</p>
    </div>
  );
};

export default memo(HomePage);
```

We just updated the component to get the component information from the `package.json` Strapi section (not the most elegant way, but this seems to be the standard in other plugins). Once we save the file, the admin panel should be refreshed automatically and our changes should be reflected when we click on our plugin link, as depicted in the following screenshot:

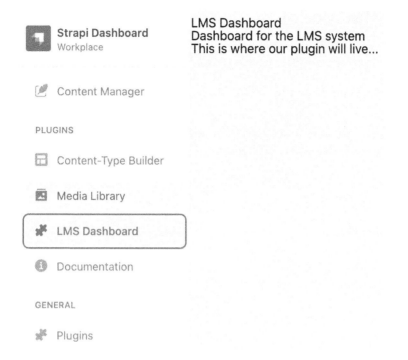

Figure 8.15: The plugin component shows in the dashboard

> **Note**
> You might have to refresh the page manually and log in again to the admin panel every time you do a change. The Strapi team is aware of this issue, and they are working on resolving it.

So far, so good. We have an entry to our plugin in the menu, and it points to our component. Now, let's start building our plugin functionality.

Building our plugin

Let's imagine we have a user requirement as follows:

As an Admin,

I want to see a list of all classes with people enrolled in them

So that I can easily have an overview of the school.

While we can get this information by navigating through each classroom, it is nice to have an overview page to see all of it at once. So, let's build a plugin to show this information.

Updating the model

First things first—let's add the relationship between the classroom and students, as we don't have this yet. This is similar to what we've done in the past, as outlined here:

1. We go to **Content-Type Builder** and add a one-to-many relationship between users and the classroom.

2. Click **Finish** then **Save** and wait for the admin panel to restart. The process is illustrated in the following screenshot:

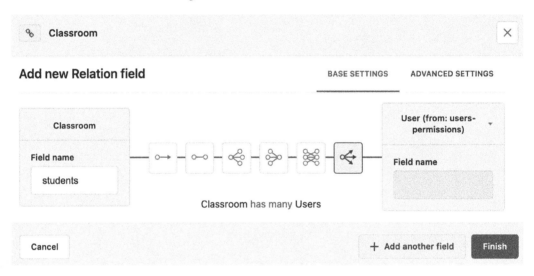

Figure 8.16: Setting up the relationship between classrooms and users

3. Now, if we go to a classroom entity in the **Content Manager** plugin, we will see that we can assign students to that classroom. Go ahead and create some users entities and assign them to classrooms, as illustrated in the following screenshot:

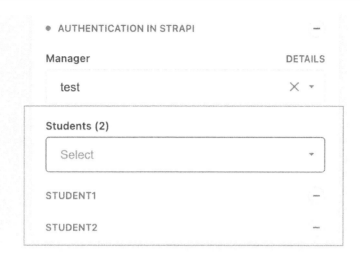

Figure 8.17: Assigning some students to classrooms

Now, let's build our plugin functionality to show all classrooms with their enrolled students in one view.

Building the Enrollments component

Now that we have a model that reflects the relationship between a classroom and its students, we will start building the **user interface** (**UI**) components to show this information, as follows:

1. Create a placeholder component to show our classroom enrollments, as follows:

```
/*
 * Enrollments (plugins/lms/admin/src/pages/Enrollments/
index.js)
 */

import React, { memo } from "react";

const Enrollments = () => {
  return (
    <div>
      <p>List of classes with enrollments</p>
    </div>
  );
};
```

```
export default memo(Enrollments);
```

2. Now, go ahead and add a route for it under the `plugins/lms/admin/src/pages/App/index.js` file. The updated module will look like this:

```
// import the module in top of the file
import Enrollments from "../Enrollments";
....
const App = () => {
  return (
    <div>
      <Switch>
        <Route path={`/plugins/${pluginId}`}
          component={HomePage} exact />
        <Route
          path={`/plugins/${pluginId}/enrollments`}
          component={Enrollments}
          exact
        />
        <Route component={NotFound} />
      </Switch>
    </div>
  );
};
```

Note that all our routes should be prefixed with `pluginId`.

3. Now, we add a link in our plugin Home page to the `Enrollments` page, as follows:

```
/*
 * HomePage (plugins/lms/admin/src/page/HomePage/index.js)
 */

import React, { memo } from "react";
import { Link } from "react-router-dom";
import pluginId from "../../pluginId";
import pluginPkg from "../../../../package.json";
```

```
const HomePage = () => {
  return (
    <div>
      <h1>{pluginPkg.strapi.name}</h1>
      <div>
        <Link to={`/plugins/${pluginId}/enrollments`}>
          Show all Enrollments
        </Link>
      </div>
    </div>
  );
};

export default memo(HomePage);
```

We use the `Link` component from the `react-router-dom` package to navigate to our new component's route. Once the admin refreshes, then we can see a link, and when this is clicked on, the placeholder for our `Enrollments` page will show.

Creating an API token

To communicate with the API from the admin panel, we will use the `axios` library to send HTTP requests. Those requests will be treated as public requests, which means the **Public** role permissions will be applied to those requests. If we want to get a list of students in the classroom, we will have to make sure that the role has access to all the content-types.

Luckily, we do not need to make all our endpoint public; instead, we can make use of a Strapi API token. This token allows us to issue requests against the API endpoint as the **Authenticated** role. Let's create a token, as follows:

1. Go to the **Settings** menu item, then click **API Tokens**.
2. Click the **Add Entry** button to create a new token.

3. Enter a name and a description for the token. Make sure to set the **Token type** value to **Read-only**, as illustrated in the following screenshot:

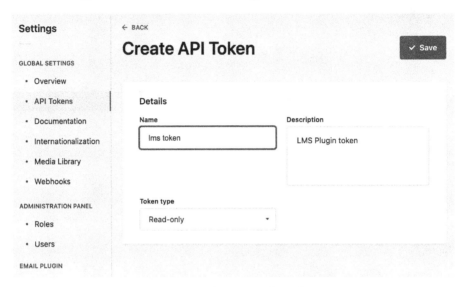

Figure 8.18: Creating an API token

4. Click the **Save** button to create the token.

5. Once you click the **Save** button, a token will be generated. Make sure to make note of the token as this is the only time you will be able to retrieve the token.

6. Define a new `API_TOKEN` variable in the `.env` file, as follows:

```
# API Token
API_TOKEN=ADD_TOKEN_HERE
```

We now have an API token and we have added it to our environment variables. However, the environment variables defined in the `.env` file are only available for the backend, not the frontend. This means our plugin will not be able to access the `process.env.API_TOKEN` variable out of the box.

7. We can use `webpack` to pass the variable to the frontend. The `webpack` configurations can be found in the `src/admin/webpack.config.js` file. Let's update this file, as follows:

```
'use strict';


/* eslint-disable no-unused-vars */
module.exports = (config, webpack) => {
```

```
    // Note: we provide webpack above so you should not
       `require` it
    // Perform customizations to webpack config
    // Important: return the modified config
    config.plugins.push(
     new webpack.DefinePlugin({
       ENV: {
        API_TOKEN: JSON.stringify(process.env.API_TOKEN),
        },
       })
     );
    return config;
  };
```

The preceding code will make the API_TOKEN variable available in the frontend using the ENV.API_TOKEN variable.

Now that we have the API_TOKEN variable available in the frontend and by extension to our plugin, let's next see how we can display a list of students on the Enrollments page.

Displaying enrollments in the plugin

Now, we can update our Enrollments component to show the classrooms and the students assigned to them in one view. Here is the code for the component:

```
/*
 * Enrollments (plugins/lms/admin/src/pages/Enrollments/index.
js)
 */

import React, { memo, useEffect, useState } from "react";
import axios from "axios";

const Enrollments = () => {
  const [classes, setClasses] = useState([]);
  useEffect(() => {
    const fetch = async () => {
      const axiosConfig = {
        headers: { authorization: `Bearer ${ENV.API_TOKEN}`
```

```
      },
    };
const { data } = await axios.get(`${strapi.backendURL}/api/
classrooms?populate=*`, axiosConfig);
      setClasses(data);
    };
    fetch();
  }, []);
  return (
    <div>
      <h1>List of classes with enrollments</h1>
      {classes.map((classroom) => {
        const { id, attributes } = item;
        const students = attributes.students.data;

        return (
        <React.Fragment key={id}>
          <div key={classroom.id}>
            <h2>{attributes.name}</h2>
            <p>{attributes.description}</p>
        <br />
            <table>
              <thead>
                <tr>
                  <th>#</th>
                  <th>name</th>
                  <th>email</th>
                </tr>
              </thead>
              {students.map((student, i) => {
                return (
                  <tr key={student.id}>
                    <td>{i + 1}</td>
                    <td>{student.attributes.username}</td>
                    <td>{student.attributes.email}</td>
                  </tr>
```

```
                    );
                })}
            </table>
        </div>
        <hr />
    </React.Fragment>
        );
    })}
    </div>
  );
};


export default memo(Enrollments);
```

The preceding code is standard React code, which should be familiar to people with exposure to React. We used the existing GET /api/classrooms API to get the information we want then display it in a custom format. We could have also used a route specific to the plugin that internally gets that data. Routes created within the plugin will be prefixed by the plugin ID. We used the axios library to make the API calls since it's included as a peer dependency for the strapi package, but we could have also used fetch or installed a custom dependency.

We have access to a global strapi object that gives us access to properties such as strapi.backendURL to get the **Uniform Resource Locator** (**URL**) of the backend so that we can call the API in our component.

Now, if we head to our plugin home page and click the **Enrollments** link, we should see the results displayed in a tabular format, with all the classes and students enrolled in them in a single view, as illustrated in the following screenshot. We can add more data and see it reflected in the report:

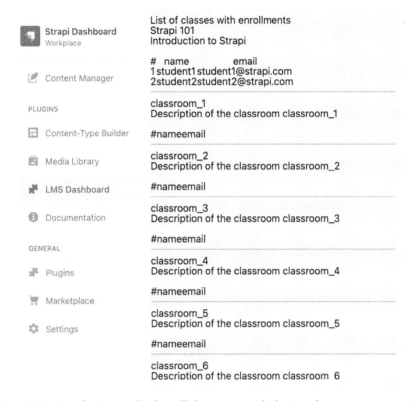

Figure 8.19: Our plugin now displays all classrooms with their students in a custom view

Using Strapi Design System

The plugin we have built is functioning correctly. However, the look and feel of it are pretty basic and dull, and it does not match the rest of the admin panel. This is where **Strapi Design System** and **UI Kit** come into play. The official page of the design system can be found here: `https://design-system.strapi.io`.

Strapi Design System is an open-source design system for Strapi plugins, as well as products and apps. It consists of pre-built reusable components that allow the user to build a consistent UI faster. The components can be found at this *Storybook* link: `https://design-system-git-develop-strapijs.vercel.app/`.

To start using it in our project, we first need to install it using the following command:

```
yarn add @strapi/design-system
```

Once we have installed it, we can then use the pre-built components to style the UI as we feel fit. We will leave this part to you to implement. The following screenshot shows an example of how the design system can be used to style the plugin interface:

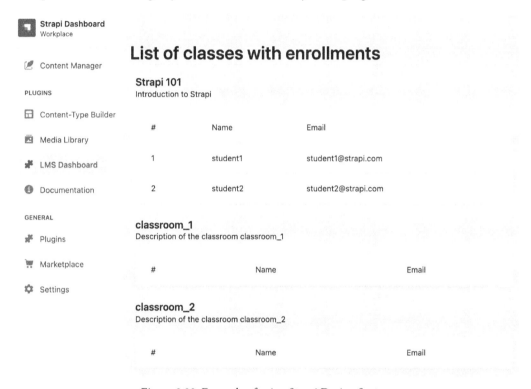

Figure 8.20: Example of using Strapi Design System

The source code in the preceding screenshot can be found in the book's GitHub repository.

Finally, before going back to running Strapi with `yarn develop`, we should run `yarn build` to rebuild the admin panel so that the admin panel is served with our new plugin.

Summary

In this chapter, we explored the Strapi plugin ecosystem. We started by explaining the concept of a plugin in Strapi and how to install and configure a plugin from the **Marketplace**. We gave an example of using the API **Documentation** plugin and saw how can it be used to generate OpenAPI specifications for our API. Then, we explained how we can find more community-based plugins from GitHub or via npm.

We then explored two popular plugins and use cases. First, we learned how to enable GraphQL for our API and how to use GraphQL Playground to test it and explore the generated schema. Next, we learned how to send emails from Strapi by using the **Email** plugin. While configuring the **Email** plugin, we also explored the concept of a provider, setting up a custom email provider for us to use in production.

Afterward, we dug deeper into the plugin architecture and created our own custom local plugin with a frontend component to show in the admin panel. This opens up endless capabilities for us to customize Strapi and its admin panel to our requirements. Finally, we had a brief introduction to Strapi Design System and UI Kit.

This is the end of the second part of the book, where we took a deeper dive into Strapi's concepts. In the next part of the book, we will be moving on to discuss Strapi in production, starting with how we can deploy our API to the cloud.

Section 3: Running Strapi in Production

Section 3 of the book focuses on preparing and running Strapi in a production environment. We will see how to use Postgres as a production database, use s3 for storing media files, keep permission in sync between environments, seed the database with sample data, and use multiple configurations for each environment. Afterward, we will learn how to deploy our API to Heroku and AWS Fargate. The final chapter is dedicated to testing the API; we will see how to write unit and integration tests for our API.

In this section, we will cover the following chapters:

- *Chapter 9, Production-Ready Applications*
- *Chapter 10, Deploying Strapi*
- *Chapter 11, Testing the Strapi API*

9
Production-Ready Applications

In this chapter, we will discuss some of the best practices, tips, and tricks, as well as strategies that we can use to run a Strapi application in a production environment. First, we will learn how to use the Strapi `bootstrap` function to seed the database. Then, we will see how to keep permissions in sync between multiple environments. Afterward, we will explain how to use **Simple Storage Service** (**S3**) to host our media files and assets. Finally, we will explain the required changes to use a database other than SQLite.

Here are the topics we will cover in this chapter:

- Create a seeder function to populate the database with required data for the API.
- Sync permissions between multiple environments.
- How to set up Strapi to save media in an S3 bucket.
- How to configure Strapi to use a Postgres database.

As we have seen so far, most of the settings and configurations such as routes, policies, and plugin settings live in code, which makes it easy to work with Strapi in different environments. However, there are a couple of things that live in the database—namely, the roles' and permissions' configurations. This means if we moved our Strapi application to a new environment or we dropped or changed the database, we would have to recreate all the roles and set up the permissions again.

Luckily, we do not need to do this manually—we can create a script that takes care of seeding the database with the initial data required by the application we are developing. Let's see how we can accomplish this task in the next section.

Seeding the database

Database seeding refers to the process of populating the database with data—this can be data required for the initial application setup or just sample data for demonstration purposes. This process is usually done when the application starts for the first time. Luckily for us, Strapi comes with a `bootstrap` lifecycle function that is executed every time the server starts. This function is located in the `src/index.js` file. Let's open this file and, for now, just print a simple hello world message, as follows:

```
Bootstrap(/*{strapi}*/){
    strapi.log.info("Hello World");
};
```

Once you have saved this simple change, the server will restart, and you should see a simple `Hello World` message printed on the screen, as follows:

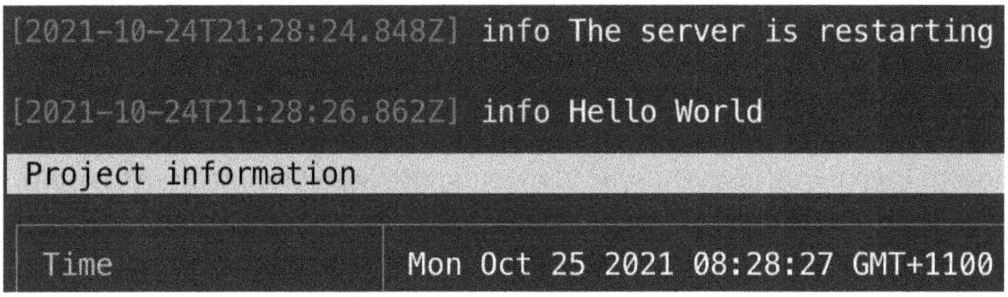

```
[2021-10-24T21:28:24.848Z] info The server is restarting

[2021-10-24T21:28:26.862Z] info Hello World

 Project information

  Time                        Mon Oct 25 2021 08:28:27 GMT+1100
```

Figure 9.1: Hello World from bootstrap function

In *Chapter 7, Authentication and Authorization in Strapi*, we created three roles for our **API** (short for **application programming interface**): **Teacher**, **Student**, and **Admin**. Let's use the `bootstrap` function to ensure that these roles are persisted in the database all the time, regardless of which database we are using or the environment we are on. In the `src/index.js` file, let's create a new `createRoleIfnotExists` function, as follows:

```
/**
 * Create role if it does not exist in the system
 * @param {*} name The role name
 * @param {*} type The role type
 * @param {*} description The role description
 */
```

```
const createRoleIfNotExist = async (name, type, description =
"") => {
    const role = await strapi.db.query("plugin::users- w
    permissions .role").findOne({ where: { type }});
    if (!role) {
      await strapi
      .db.query("plugin::users-permissions .role").create({ name,
      type, description });
      strapi.log.debug('Created role ${name}');
    }
};
```

The `createRoleIfnotExists` function takes three parameters: `role name`, `role type`, and—optionally—`description`. To avoid having duplicate roles, we start by making sure that the role does not exist in the system.

> **Note**
>
> Since we are running the query against the `user-permissions` plugin, we prefixed the model name with `plugin::` rather than `api::`, and then we specified the plugin name and the model name.

After we have ensured that the role does not exist in the database, we simply call the `create` method to create the role and print our success message to the console. Let's use this method to create the three roles needed for our API. We can remove the `Hello World` message from the `bootstrap` function and replace it with the following code:

```
async bootstrap(/*{strapi}*/){
  await createRoleIfNotExist("Student", "student", "Student
  role");
  await createRoleIfNotExist("Teacher", "teacher", "Teacher
  role");
  await createRoleIfNotExist("Admin", "admin", "App admin
  role");
};
```

Make sure to add the `async` keyword to the `bootstrap` function signature.

To see the seeding function in action, let's delete one of the roles we have already created previously, as follows:

1. Click **Settings** in Strapi's left sidebar menu.
2. Select **Roles** under the **User & Permissions Plugin** section.
3. Click the **Delete** button next to the **Admin** role and confirm you want to delete the role.

Now that we have a role missing, let's restart the server to confirm the data will be seeded. From the active terminal window, hit *Ctrl + C* to stop the server, then issue the `yarn develop` command. Notice in the following screenshot that as the server starts, we can see a `Created role Admin` message:

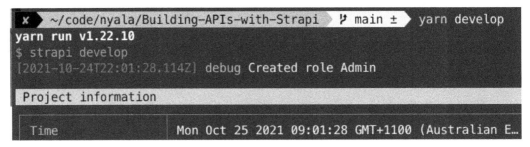

Figure 9.2: Admin role created on startup

If you refresh the **Role** page, you should see the **Admin** role is back in the role list.

Now that we can seed roles and make sure they are always in the database, let's see how we can persist the permissions of each role in the next section.

Keeping permissions in sync

The permissions associated with each role are persisted in the database. This means we will need to make sure that the permissions we created while developing the application are kept in sync when we move to a new environment or when we seed the database with a new role. For example, in the previous section, we dropped and recreated the **Admin** role; by doing so, we lost all the previous permissions we configured for this role.

To keep the permissions in sync, we will again make use of the bootstrap function. We can add a new function that will enable permissions for a specific role when the server starts up. Let's get coding, as follows:

1. In the src/index.js file, let's create a new enablePermission function that accepts three parameters, as follows: the roleType parameter for the role for which we want to enable the permission, the controller parameter, and the action parameter for the action we want to enable. The code is illustrated in the following snippet:

```
/**
 * Enable action on a controller for a specific role
 * @param {*} type The role type
 * @param {*} apiName The name of the api where the
controller lives
 * @param {*} controller The controller where the action
lives
 * @param {*} action The action itself
 */
const enablePermission = async (type, apiName,
controller, action) => {
};
```

2. First, we will get the role from the database and check the permissions associated with the role, as follows:

```
...
try {
  const actionId =
'api::${apiName}.${controller}.${action}'
  // Get the permissions associated with the role
  const rolePermission = role.permissions.find(permission
=> permission.action === actionId);

} catch (e) {
    strapi.log.error(
    'Bootstrap script: Could not update settings.
```

```
    ${controller} -    ${action}'
  );


}
```

3. Then, we will check if there are permissions associated with the role. If there are no permissions, it means this is the first time we are running the server, so we create those permissions, as follows:

```
...
if (!rolePermission) {
  // Permission not created yet (first time starting
  // the server)
  strapi.db.query("plugin::users-
    permissions.permission").create({
      data : {
        action: actionId,
        role: role.id
      }
    });
  }
...
```

Here is the full code for the enablePermission function:

```
/**
 * Enable action on a controller for a specific role
 * @param {*} type The role type
 * @param {*} apiName The name of the api where the
controller lives
 * @param {*} controller The controller where the action
lives
 * @param {*} action The action itself
 */
const enablePermission = async (type, apiName,
controller, action) => {
```

```
try {
  // Get the role entity
  const role = await strapi.db.query("plugin::users-
  permissions.role")
      .findOne({
      where: { type },
      populate: ["permissions"]
    });

  const actionId =
  'api::${apiName}.${controller}.${action}'
  // Get the permissions associated with the role
  const rolePermission = role.permissions.find(permission
  => permission.action === actionId);

  if (!rolePermission) {
    // Permission not created yet (first time starting
    // the server)
    strapi.db.query("plugin::users-
    permissions.permission").create({
     data : {
      action: actionId,
      role: role.id
     }
    });
  }
} catch (e) {
  strapi.log.error(
  'Bootstrap script: Could not update settings.
   ${controller} -   ${action}'
  );
}
};
```

4. To test this new function, let's update the `bootstrap` function as follows to enable all actions on the `Classroom` controller:

```
async bootstrap(/*{strapi}*/){
  await createRoleIfNotExist("Student", "student",
  "Student role");
  await createRoleIfNotExist("Teacher", "teacher",
  "Teacher role");
  await createRoleIfNotExist("Admin", "admin", "App
  admin role");

  // Admin Role Classroom permissions
  await enablePermission("admin",
  "classroom","classroom", "create");
  await enablePermission("admin",
  "classroom","classroom", "find");
  await enablePermission("admin",
  "classroom","classroom", "findOne");
  await enablePermission("admin",
  "classroom","classroom", "findTutorials");
  await enablePermission("admin",
  "classroom","classroom", "create");
  await enablePermission("admin",
  "classroom","classroom", "update");
  await enablePermission("admin",
  "classroom","classroom", "delete");
};
```

5. Save the changes to restart the server. Once the server restarts, examine the **Admin** role permissions. You should see that all classroom permissions are enabled, as illustrated in the following screenshot:

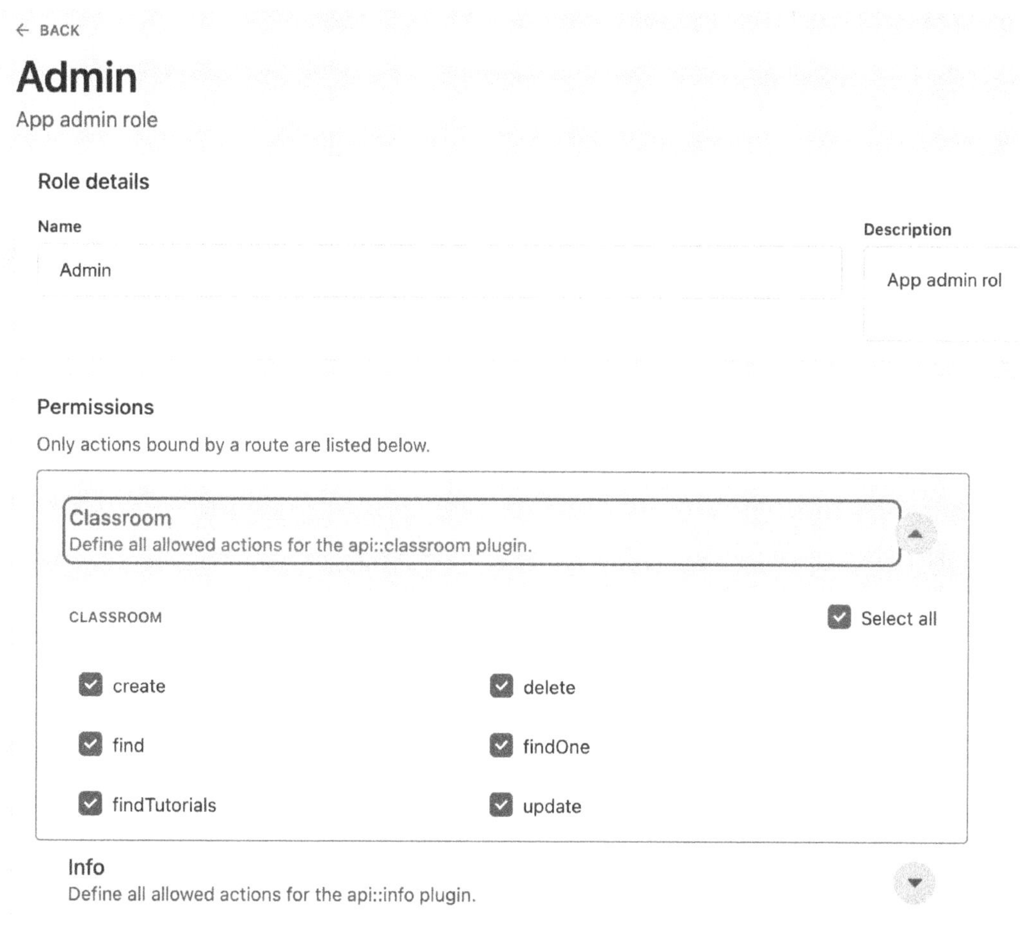

← BACK

Admin
App admin role

Role details

Name

Admin

Description

App admin rol

Permissions

Only actions bound by a route are listed below.

Classroom
Define all allowed actions for the api::classroom plugin. ▲

CLASSROOM   ☑ Select all

☑ create   ☑ delete

☑ find   ☑ findOne

☑ findTutorials   ☑ update

Info
Define all allowed actions for the api::info plugin. ▼

Figure 9.3: Admin role permissions enabled

This example enables all the permissions for the **Admin** role on the **Classroom** content-type. We can do the same for the other content-types as well as other roles. However, we have left out this part for you to do as an exercise.

So far, we have seen how to seed our database as well as keep permissions in sync. Both are important steps when we are preparing our application for production deployment. Another important step is to make sure we are serving media files and assets from a reliable source. In the next section, we will discuss how to use Amazon S3 to serve our images and media files.

Serving media from an S3 bucket

Amazon S3 is a reliable and secure cloud-based storage service offered by **Amazon Web Services** (**AWS**). We are going to configure our API to use S3 to store media files rather than maintaining them on a server physical disk.

Because S3 is a cloud-based solution there are a few advantages of using it over the local hard disk, one of which is having high scalability, as we are not going to be limited by the physical disk space. Another advantage is having better **disaster recovery** (**DR**) capability and availability. For example, with local disk storage, if the hard disk crashes, then our API users will not be able to access their files. However, the risk is much lower with Amazon S3.

> **Note**
>
> Creating an AWS account is beyond the scope of this book. We are going to assume that you already have your account up and ready. If you do not have an account, head to `https://aws.amazon.com/free` and create an account. AWS offers a free 1-year trial when you sign up.

Before we configure Strapi to use S3, we will need to create an S3 bucket first, so let's do that now. Proceed as follows:

1. In the AWS console, navigate to **Amazon S3**.
2. Click the **Create bucket** button.
3. Enter a bucket name and choose the region closest to you.
4. In the **Access** section, uncheck **Block all public access**.

    A warning box might appear, to confirm that the bucket will be publicly accessible. Check this to make it accessible to the public.

5. We can leave the remaining settings as they are. Scroll down and click the **Create bucket** button, as illustrated in the following screenshot:

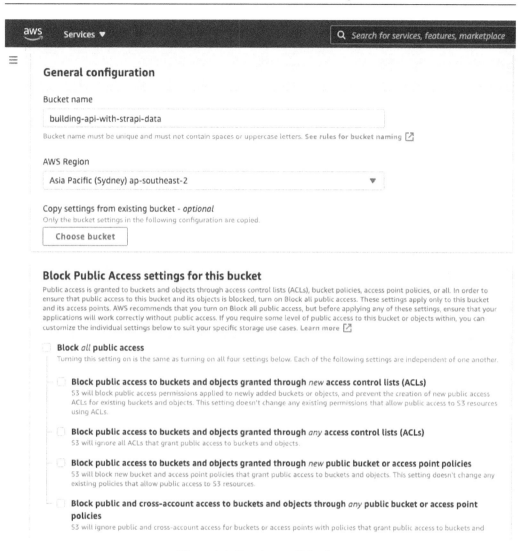

Figure 9.4: Creating an S3 bucket

The next step is to install a plugin that allows us to upload media to S3. Let's go ahead and install the plugin with the `yarn add @strapi/provider-upload-aws-s3` command.

Once the plugin is installed, we can configure Strapi to use it as the default provider when uploading images and media files. Similar to what we did in *Chapter 8, Using and Building Plugins*, with the **Email** plugin, we will configure the upload provider to be used in a production environment only. Open the `config/env/production/plugins.js` file and add the following code to it:

```
upload: {
    config: {
        provider: 'aws-s3',
        providerOptions: {
          accessKeyId: env('AWS_ACCESS_KEY_ID'),
          secretAccessKey: env('AWS_ACCESS_SECRET'),
          region: env('AWS_REGION'),
          params: {
            Bucket: env('AWS_BUCKET'),
          },
        },
    },
},
```

There are four environment variables needed. `AWS_ACCESS_KEY_ID` and `AWS_ACCESS_SECRET` can be obtained from your AWS account, `AWS_REGION` is the region where we created the bucket, and `AWS_BUCKET` is the name of the bucket we created earlier. The `.env` file should look similar to this:

```
AWS_ACCESS_KEY_ID=ADD_AWS_KEY_HERE
AWS_ACCESS_SECRET=ADD_AWS_SECRET_HERE
AWS_REGION=ap-southeast-2
AWS_BUCKET=building-api-with-strapi-data
```

To test out the plugin, let's start the server with the `NODE_ENV` flag set to `production` and then proceed as follows:

1. Make sure the development server is not running.

2. Issue the `NODE_ENV=production yarn develop` command to start the server.

3. In the Strapi admin panel, click the **Media Library** menu item from the left-side panel.

4. Click the **Upload assets** button to upload a new image.

5. Once the image is uploaded, click on it, then click the **Copy link** button.

6. Paste the link into a new tab—notice that the link is an AWS link. You can also check the bucket we created earlier, and you will see the image file there. In the following screenshot, you can see an example of an S3 bucket after uploading media:

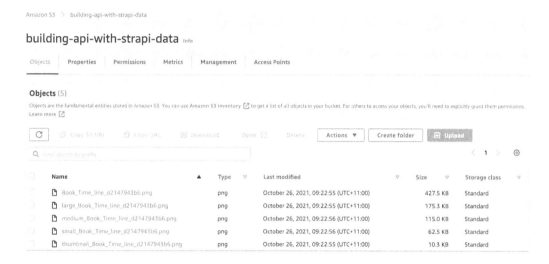

Figure 9.5: Example of an S3 bucket after uploading media

> **Troubleshooting**
>
> If you are having issues with the upload, make sure you are using the correct access key **identifier** (**ID**) and secret, and also ensure that these permissions are granted to the user in the AWS **Identity and Access Management** (**IAM**) policy settings: `"s3:PutObject"`, `"s3:GetObject"`, `"s3:ListBucket"`, `"s3:DeleteObject"`, `"s3:PutObjectAcl"`.

Now that we have set up the `upload` function, let's shift our attention to the next item in preparing our application for production—the database. So far, we have been using an **SQLite database**, which is fine for development environments or low web traffic systems. However, when we are working in a production environment with a large volume of traffic, we will probably want to use a different database system. In the next section, we will discuss how can we set up Strapi with **PostgreSQL**.

Using PostgreSQL

You might recall from *Chapter 1*, *An Introduction to Strapi*, that Strapi is database-agnostic, meaning that it can work with different database systems. We will configure Strapi to use a PostgreSQL database in production and keep SQLite for development. Let's get started, as follows:

1. The first thing to do is to install the PostgreSQL client.

2. From the terminal, navigate to the project root and run the `yarn add pg` command. This will install the PostgreSQL Node.js client to the project.

3. Create a `config/env/production/database.js` file. This file will be used to configure the database in the production environment.

> **Remember**
>
> Strapi uses the `config/database.js` file to load the database configurations. However, since we want to use PostgreSQL in the production environment only, we used the `config/env/production/database.js` file instead.

4. Add the following content:

```
module.exports = ({ env }) => ({
  connection: {
    client: 'postgres',
    connection: {
      host: env('DATABASE_HOST', 'localhost'),
      port: env.int('DATABASE_PORT', 5432),
      database: env('DATABASE_NAME', 'strapi'),
      user: env('DATABASE_USERNAME', 'strapi'),
      password: env('DATABASE_PASSWORD', 'password'),
      ssl: env.bool('DATABASE_SSL', false),
    },
  },
});
```

The following table explains each of the preceding configurations used:

Key	Description	Sample Value
`client`	The name of the client to use. Possible values are `postgres`, `mysql`, or `sqlite`.	`postgres`

`-port`
 	
`-database`
 	
`-user`
 	
`-password`
	Database connection string values.	`The values depend on the database configuration`

`ssl`	A flag to indicate whether to use **Secure Sockets Layer** (**SSL**) when connecting to the database.	`False`

Our API is now ready to work with PostgreSQL when we deploy it to production. However, it is always good to test settings and configuration first to avoid any surprises later on.

Testing PostgreSQL locally

To test the database configurations locally, we will need to have PostgreSQL installed. We will be using Docker to run PostgreSQL. Let's get started with the following steps:

1. At the root of the project, let's create a new file called `docker-compose-dev.yml`. This file will hold the database configuration.

2. Open the file and add the following code to it:

```
version: "3.1"
services:
  database:
    container_name: strapi_api
    image: postgres
    ports:
      - 5432:5432
    environment:
      POSTGRES_DB: strapi
      POSTGRES_USER: strapi
```

```
        POSTGRES_PASSWORD: password
     volumes:
        - strapi_db:/var/lib/postgresql/data
  volumes:
     strapi_db: {}
```

3. Save the changes to the file. This docker-compose-dev.yml file will create a service called database to run the postgres database. It will create a database called strapi with a username of strapi and a password of password.

4. Next, let's add helper scripts to our project to allow us to start and stop the database. Open the package.json file and locate the scripts section.

5. Add the following scripts at the end of the scripts section:

```
"db:start": "docker compose -f docker-compose-dev.yml up
-d",
"db:stop": "docker compose -f docker-compose-dev.yml
down",
        "db:logs": "docker logs -f strapi_api"
```

6. Save the changes and let's test out our scripts.

7. Run the db:start command as follows: yarn db:start. If the command is executed successfully, you will see a message on the console indicating that a container named strapi_api has started.

> **Note**
> Running this command for the first time might take a minute or so as Docker will pull the Postgres database image.

Once we have our PostgreSQL database up and running, let's start the server with the NODE_ENV flag set to production. Issue the NODE_ENV=production yarn develop command to start the server.

Since this is a new database, we will be asked to create a super admin user. Create an admin account and log in to the dashboard. Everything should be exactly the same; also, the database should have been seeded with the roles.

You can shut down the server now, and make sure to stop the database by running the yarn db:stop command.

Our API is now ready for production deployment. This will be the topic of our next chapter.

Summary

In this chapter, we explored best practices and strategies we can use to prepare and run our application in a production environment. We started with database seeding and saw how we can seed the database to have the data required for the initial application setup. The database seeder we created always initializes the database with the required user roles for the API.

Next, we discussed how can we keep permissions in sync across multiple environments, and we created a helper function that will do this job for us. After that, we moved on to media and assets and configured Strapi to use Amazon S3 to save media files instead of saving them on the local hard drive.

Finally, we prepared our API to use the PostgreSQL database in production, and we also tested it out locally with the help of Docker.

In the next chapter, we will explore how to deploy our API to a production environment and we will discuss two deployment strategies: deployment to a **software-as-a-service** (**SaaS**) provider and deployment as a Docker container.

10
Deploying Strapi

We have been adding functionality to our API in the past few chapters and we are finally ready to deploy it to a production environment where it can see the light. In this chapter, we will learn two of the most popular strategies or methods to deploy a Strapi application. First, we will learn how to deploy the API to a **Platform as a Service** (**PaaS**). We will use the Heroku platform for this example. Next, we will learn how to deploy the API as a Docker container to AWS Fargate. By the end of this chapter, you will understand the difference between two of the most popular platforms, Heroku and AWS, and be able to deploy your code to secure, reliable platforms.

The topics we will cover in this chapter are the following:

- Deploying a Strapi app to Heroku
- Deploying a Strapi App to AWS Fargate

Deploying to Heroku

Heroku is a popular **PaaS** that enables developers to build, run, and operate applications in the cloud. We are going to use Heroku to deploy our Strapi application. There are two simple requirements:

- Have a Heroku account set up and the Heroku CLI installed.
- You need to have Git installed locally.

> **Important Note**
>
> Creating Heroku accounts is beyond the scope of this book, so we are going to assume you already have an account set up. If you do not have an account yet, you can take a pause here, head to the Heroku website, create your free account, and configure the Heroku CLI for your operating system.

The first thing we need to do is to create an application on Heroku. There are two ways to do so. The first one is using the Heroku dashboard, and the second one is using the Heroku CLI. We are going to use the CLI, so let's get started:

1. Launch your terminal and log in to Heroku using the login command `heroku login`. Press any key to open your browser and log in to Heroku.

2. Return to the terminal after you log in. You should see login success information on the terminal.

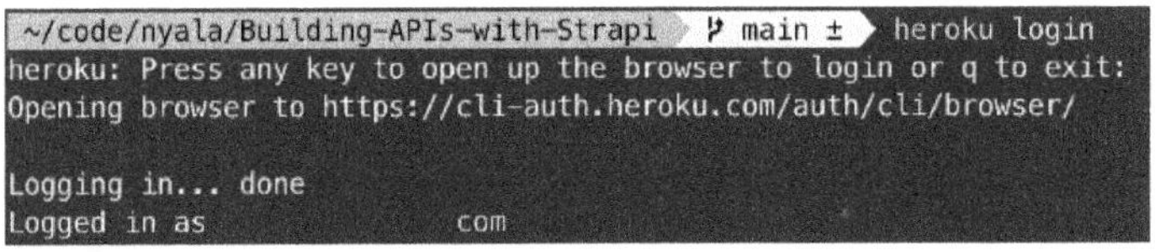

```
~/code/nyala/Building-APIs-with-Strapi    ⌥ main ±    heroku login
heroku: Press any key to open up the browser to login or q to exit:
Opening browser to https://cli-auth.heroku.com/auth/cli/browser/

Logging in... done
Logged in as                    com
```

Figure 10.1: Heroku login

3. Make sure you are in the root folder of the project. This is important because when we create a Heroku app in the next step, it will also update the project's **Git** configuration adding a new **Git remote**.

4. Next, we will use the `create` command to create a new app. We will give our Heroku app a name and we will specify the region as well. The command will look like this:

```
heroku create strapi-lms-api --region eu
```

Note that you can use the `create` command without any parameters. If we did not specify a name, then Heroku will use a randomly generated name and URL for our app. The `region` parameter is completely optional and you can choose between `us` and `eu`. We chose `eu` as it's the closest `region` to us.

> **Note**
>
> The app name must be unique across all Heroku apps. The `strapi-lms-api` name we chose here will not work for you as it has been taken already now.

5. Once you issue the `create` command, Heroku will reply with a success message if the app was created, and two URLs. The first URL is the API URL. We can use this one to access our API. The second URL is a Git repository URL on the Heroku servers. We will use this URL to deploy our API to Heroku.

```
x  ~/code/nyala/Building-APIs-with-Strapi  ⅄ main ±  heroku create strapi-lms-api --region eu
Creating ● strapi-lms-api... done, region is eu
https://strapi-lms-api.herokuapp.com/ | https://git.heroku.com/strapi-lms-api.git
```

Figure 10.2: Creating a new Heroku app

If you clicked on the first URL to open the app. You will see the default Heroku welcome message. This is fine for now. It will change later on when we deploy the API.

6. Make a note of the application URL as we will need it in a later step.

7. The next step is to install the Heroku Postgres add-on. The Postgres add-on will allow us to use the Postgres database. Let's use the `addons` command as follows:

```
heroku addons:create heroku-postgresql:hobby-dev
```

The `hobby-dev` suffix in the `addons` command will set up the free Postgres tier for us, which is more than enough for our demo purposes. However, it is recommended to use one of the paid plans when you are deploying your API to production.

You should see a message confirming that the database was created successfully in the terminal as well as the database name and a link to view the database documentation.

```
~/code/nyala/Building-APIs-with-Strapi  ⅄ main ±  heroku addons:create heroku-postgresql:hobby-dev
Creating heroku-postgresql:hobby-dev on ● strapi-lms-api... free
Database has been created and is available
 ! This database is empty. If upgrading, you can transfer
 ! data from another database with pg:copy
Created postgresql-aerodynamic-76170 as DATABASE_URL
Use heroku addons:docs heroku-postgresql to view documentation
```

Figure 10.3: Creating a database on Heroku

8. The previous command will also add an environment variable called DATABASE_

URL to our Heroku app. This variable holds the full connection string to our newly created database. The format is as follows:

```
postgress://USER_NAME:PASSWORD@SERVER_IP:PORT/DATABASE_
NAME
```

9. If you open the database configuration file `config/env/production/database.js`, you will notice that we need each of the database connection configurations (`host`, `user`, `password`, `port`, and `database` name) as separate environment variables. So we can either copy the needed values from the database URL one by one or use a library that will parse the connection string for us. Let's do the latter and install a package called `pg-connection-string`. From the terminal, issue this command:

```
yarn add pg-connection-string
```

10. Once you have installed the package, let's update the database configuration in `config/env/production/database.js` as follows:

```
const { parse } = require('pg-connection-string');
const dbConfig = parse(process.env.DATABASE_URL);

module.exports = ({ env }) => ({
  connection: {
    client: 'postgres',
    connection: {
      host: dbConfig.host,
      port: dbConfig.port,
      database: dbConfig.database,
      user: dbConfig.user,
      password: dbConfig.password,
      ssl: env.bool('DATABASE_SSL', false),
    },
  },
});
```

We started by requiring the parse function from the pg-connection-string
package. Then, we used the parse function with the DATABASE_URL environment
variable to parse the database connection string and set the values in the dbConfig
variable. Finally, we updated all the database connection parameters to read the
values from the dbConfig variable.

11. We are going to make one more change to the database configuration. We will add
the rejectUnauthorized option to the database configuration. This is because
Heroku does not support client-side certificate validation to its Postgres database
unless we are using a Private or Shield Heroku Postgres database. The change should
look as follows:

```
const { parse } = require('pg-connection-string');
const dbConfig = parse(process.env.DATABASE_URL);

module.exports = ({ env }) => ({
  connection: {
    client: 'postgres',
    connection: {
      host: dbConfig.host,
      port: dbConfig.port,
      database: dbConfig.database,
      user: dbConfig.user,
      password: dbConfig.password,
      ssl: {
        RejectUnauthorized: false
      },
    },
  },
});
```

12. The final change we need before deploying our API is to let Strapi know our public Heroku domain. This should be the URL we got in step 5 after we used the `create` command. Create a new server configuration file, `config/env/production/server.js`, and add the following:

```
module.exports = ({ env }) => ({
    url: env("API_HEROKU_URL"),
});
```

This will let Strapi use the URL defined in the `API_HEROKU_URL` variable as the server URL.

13. Let's configure the `API_HEROKU_URL` variable using the `config:set` command as follows:

```
heroku config:set API_HEROKU_URL=YOUR_HEROKU_URL_HERE
```

14. Make sure to replace `YOUR_HEROKU_URL_HERE` with the actual URL for your Heroku app.

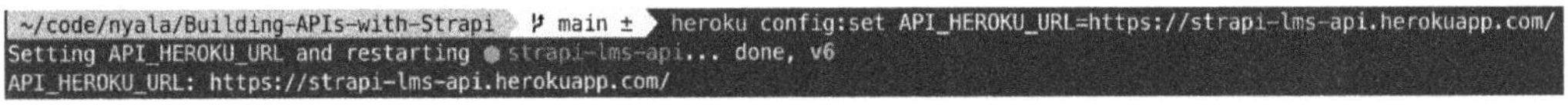

```
~/code/nyala/Building-APIs-with-Strapi  ⎇ main ±  heroku config:set API_HEROKU_URL=https://strapi-lms-api.herokuapp.com/
Setting API_HEROKU_URL and restarting ● strapi-lms-api... done, v6
API_HEROKU_URL: https://strapi-lms-api.herokuapp.com/
```

Figure 10.4: Setting a Heroku environment variable example

15. Finally, let's set all the required environment variables for our API. We will use the `config:set` command again to set the required environment variables for our API. The following is a list of all `heroku config:set` commands:

```
heroku config:set NODE_ENV=production
heroku config:set AWS_ACCESS_KEY_ID=VALUE_HERE
heroku config:set AWS_ACCESS_SECRET=VALUE_HERE
heroku config:set AWS_BUCKET=VALUE_HERE
heroku config:set AWS_REGION=VALUE_HERE
heroku config:set MAILGUN_API_KEY=VALUE_HERE
heroku config:set MAILGUN_DOMAIN=VALUE_HERE
```

16. We can check what environment variables are currently set using the `heroku` `config` command. This can help you verify that all the required variables have been set and we are ready for deployment.

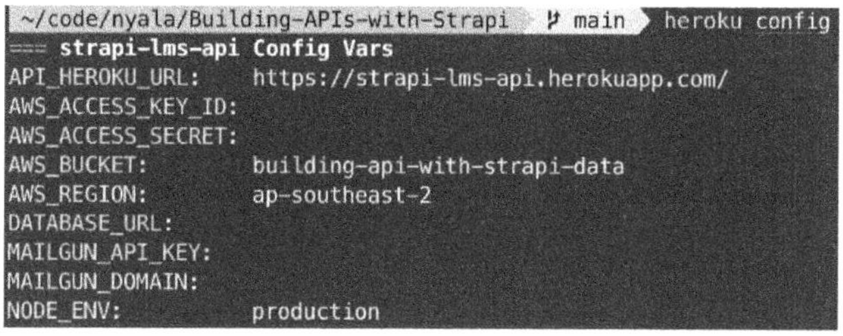

```
~/code/nyala/Building-APIs-with-Strapi    main   heroku config
=== strapi-lms-api Config Vars
API_HEROKU_URL:         https://strapi-lms-api.herokuapp.com/
AWS_ACCESS_KEY_ID:
AWS_ACCESS_SECRET:
AWS_BUCKET:             building-api-with-strapi-data
AWS_REGION:             ap-southeast-2
DATABASE_URL:
MAILGUN_API_KEY:
MAILGUN_DOMAIN:
NODE_ENV:               production
```

Figure 10.5: Displaying environment variables on the Heroku app

17. To deploy the API to Heroku, all we need to do is simply push the changes to the `heroku` Git remote using the `git push` command: `git push heroku MAIN_BRANCH_NAME` where `MAIN_BRANCH_NAME` is the name of your main or master branch:

```
git add .
git commit -m "Deploying to heroku"
git push heroku main
```

18. Once you have pushed the changes, you should see a build message in the terminal. It will take a couple of minutes for Heroku to build and deploy the app.

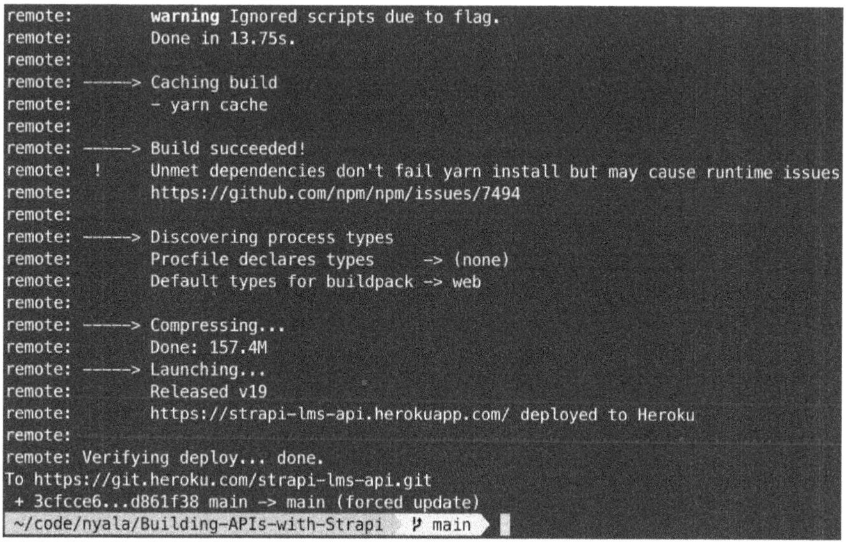

```
remote:        warning Ignored scripts due to flag.
remote:        Done in 13.75s.
remote:
remote: -----> Caching build
remote:        - yarn cache
remote:
remote: -----> Build succeeded!
remote:  !     Unmet dependencies don't fail yarn install but may cause runtime issues
remote:        https://github.com/npm/npm/issues/7494
remote:
remote: -----> Discovering process types
remote:        Procfile declares types      -> (none)
remote:        Default types for buildpack  -> web
remote:
remote: -----> Compressing...
remote:        Done: 157.4M
remote: -----> Launching...
remote:        Released v19
remote:        https://strapi-lms-api.herokuapp.com/ deployed to Heroku
remote:
remote: Verifying deploy... done.
To https://git.heroku.com/strapi-lms-api.git
 + 3cfcce6...d861f38 main -> main (forced update)
~/code/nyala/Building-APIs-with-Strapi    main 
```

Figure 10.6: Deploying to Heroku success message

19. If you open the Heroku app URL, you should see the page is now displaying the Strapi main page, and the environment label is showing **PRODUCTION**.

Figure 10.7: Heroku URL and Strapi PRODUCTION

Our API was successfully deployed to Heroku. You can navigate to the **/admin** page now and log in to the dashboard. Remember that you will be asked to create the **Super Admin** user the first time you log in to the dashboard.

Now that we have learned how to deploy a Strapi app to a PaaS platform, let's discuss how can we use another strategy to deploy to a serverless infrastructure using AWS Fargate and Docker.

Deploying to AWS Fargate

AWS Fargate is a service from AWS that allows users to run containers on the AWS platform without the need to manage the underlying infrastructure. We will learn how to deploy our Strapi API as a Docker container to AWS Fargate. The requirements are as follows:

- Have an AWS account and set up the AWS CLI

- Have the Docker CLI installed locally

There are a few steps involved in deploying our API to AWS Fargate. Here is an overview of all the steps. First, we will create a containerized version of our API using Docker. Next, we will use **Amazon Elastic Container Registry** (**Amazon ECR**) to create a Docker repository. We will then push our Strapi API Docker image to this repository. Finally, we will create a Fargate cluster that will use the Docker repository to run our API.

Step 1 – Creating a Docker image for our API

The first step is to create a Docker image for our API, so let's get started:

1. Create a new file called `Dockerfile` at the root of our project.

2. Add the following to the newly created `Dockerfile`:

```
FROM node:16-alpine
# Our working directory
WORKDIR /app

# Set node env
ENV NODE_ENV=production
# Copy the required files
COPY package.json .
COPY yarn.lock .
COPY favicon.ico .
COPY api/ api/
COPY components/ components/
COPY config/ config/
COPY extensions/ extensions/
COPY plugins/ plugins/
COPY public/robots.txt public/

# Install dependencies and run build
RUN yarn --production --frozen-lockfile
RUN yarn build

EXPOSE 1337

CMD ["yarn", "start"]
```

We started the `Dockerfile` by defining the base image `node:16-alpine`. This is a minimal Node 16 image running on Alpine Linux. We used this version to match the node requirements specified in the `package.json` file.

Then we specified the working directory of our container, and we set the NODE_ENV variable to production. After that, we started copying the required files from our project to the container.

Next, we installed the dependencies. Note that we specified the -frozen-lockfile flag to indicate that we do not want to generate a yarn.lock file and use the current lockfile instead.

Finally, we exposed port 1337 from the container to be accessible to the outside world and specified that we want to run the yarn start command when the container starts.

3. Next, let's add a .dockerignore file to specify folders and files that should be ignored by the Docker client when building an image. Create a new file called .dockerignore in the project root.

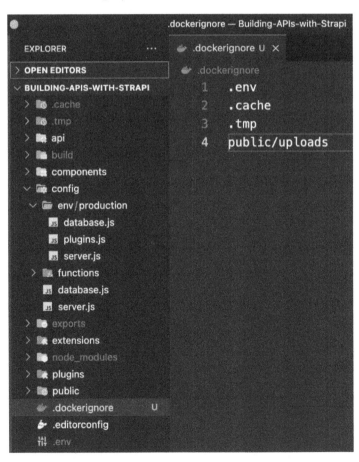

Figure 10.8: .dockerignore file

4. We are ready to build the Docker image, but before we do that we will make a minor change to the `config/env/production/database.js` file. Open the file and change the parse line as follows:

```
const { parse } = require("pg-connection-string");
const dbConfig = parse(process.env.DATABASE_URL || "");
...
```

The reason for this change is that if we were to build the Docker image now, the `pg-connection-string` parser we used before would expect the DATABASE_ URL environment variable to be present during the `yarn build` step. If it is not present, then the `yarn build` command will fail, which will cause the Docker image build to fail.

5. Run the `docker build` command to build the image. We will also tag the image, giving it the name `strapi-lms`. The `build` command is as follows:

```
docker build -t strapi-lms .
```

Note the dot at the end of the command is to let Docker know that the `Dockerfile` is located in the current directory.

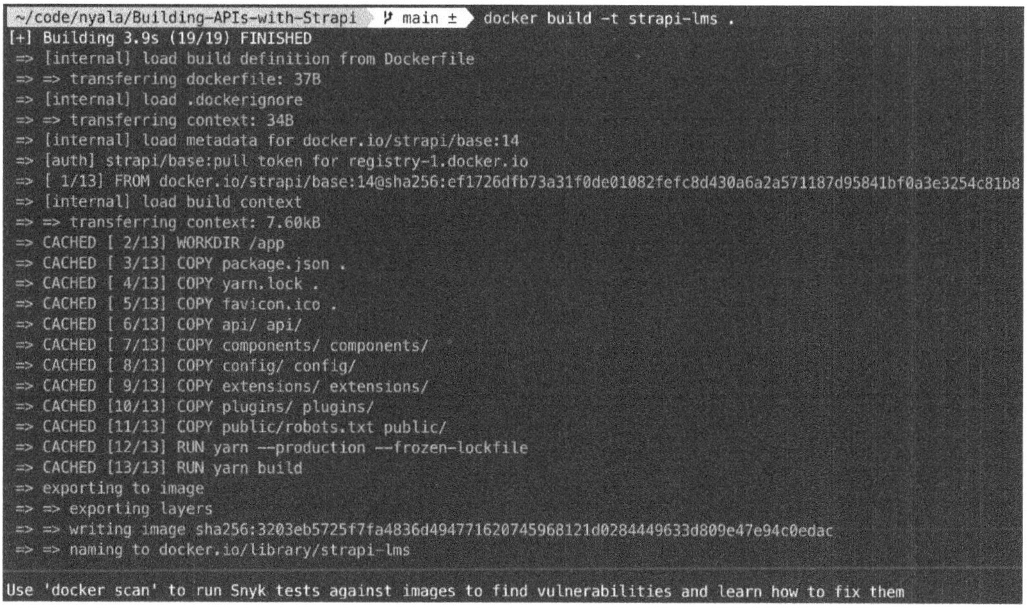

```
~/code/nyala/Building-APIs-with-Strapi  ⑂ main ±  docker build -t strapi-lms .
[+] Building 3.9s (19/19) FINISHED
 => [internal] load build definition from Dockerfile
 => => transferring dockerfile: 37B
 => [internal] load .dockerignore
 => => transferring context: 34B
 => [internal] load metadata for docker.io/strapi/base:14
 => [auth] strapi/base:pull token for registry-1.docker.io
 => [ 1/13] FROM docker.io/strapi/base:14@sha256:ef1726dfb73a31f0de01082fefc8d430a6a2a571187d95841bf0a3e3254c81b8
 => [internal] load build context
 => => transferring context: 7.60kB
 => CACHED [ 2/13] WORKDIR /app
 => CACHED [ 3/13] COPY package.json .
 => CACHED [ 4/13] COPY yarn.lock .
 => CACHED [ 5/13] COPY favicon.ico .
 => CACHED [ 6/13] COPY api/ api/
 => CACHED [ 7/13] COPY components/ components/
 => CACHED [ 8/13] COPY config/ config/
 => CACHED [ 9/13] COPY extensions/ extensions/
 => CACHED [10/13] COPY plugins/ plugins/
 => CACHED [11/13] COPY public/robots.txt public/
 => CACHED [12/13] RUN yarn --production --frozen-lockfile
 => CACHED [13/13] RUN yarn build
 => exporting to image
 => => exporting layers
 => => writing image sha256:3203eb5725f7fa4836d494771620745968121d0284449633d809e47e94c0edac
 => => naming to docker.io/library/strapi-lms
Use 'docker scan' to run Snyk tests against images to find vulnerabilities and learn how to fix them
```

Figure 10.9: Building a Docker image

The `build` command will take a couple of minutes to finish. Once the build is complete, you can run the `docker images` command to list the Docker images you have. You should see a Docker image with the name `strapi-lms`.

Step 2 – Creating a Docker repository on AWS ECR

Now that we have a Docker image, let's create a Docker repository on AWS ECR to push our image to:

1. Log in to the AWS console and navigate to the **Amazon Elastic Container Services** page (you can search for `fargate` in the search bar and then select **Elastic Container Service**).

2. From the left-hand side menu, click on **Repositories**, then click the **Create repository** button to create a new ECR repository.

3. Keep the visibility settings as **Private** and choose a name for the repository. We will use `strapi-lms` as the repository name.

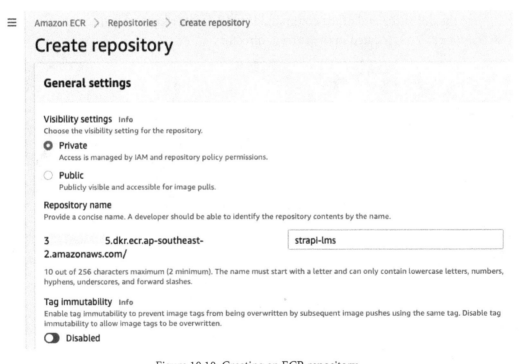

Figure 10.10: Creating an ECR repository

4. Leave the rest of the settings as is, scroll down, and click the **Create repository** button.

5. Once the repository is created, you will be redirected to the repository list page. Make note of the repository **URI** as we will need it in the next step.

6. Since this is a private repository, we will need to log in to it first. The login command is this:

    ```
    aws ecr get-login-password --region REGION_NAME_HERE |
    docker login --username AWS --password-stdin REPO_URL_
    HERE
    ```

    Make sure to replace REGION_NAME_HERE with the region name you are working on and REPO_URL_HERE with the URI we got from the previous step.

    Figure 10.11: Log in to AWS ECR

    It will take a few seconds, then you should see the **Login Succeeded** message in the terminal.

7. Next, let's tag the Docker image we created earlier with the repository URI so we can push it to AWS. This is the command:

    ```
    docker tag strapi-lms:latest REPO_URL_HERE:latest
    ```

    Make sure to replace REPO_URL_HERE with your repository URI.

    > TIP
    >
    > We are using latest as our Docker image tag. However, you might want to use a better tag for the images along with the latest tag to indicate the image version. One strategy to consider is using build numbers from a CI/CD tool or a commit SHA as an image tag.

8. Once we've tagged the Docker image, we can push it to AWS ECR using the `docker push` command, so let's do that by running this command:

```
docker push REPO_URL_HERE:latest
```

Make sure to replace REPO_URL_HERE with your repository URL.

```
x  ~/code/nyala/Building-APIs-with-Strapi  ⑂ main ±  docker tag strapi-lms:latest 3        .dkr.ecr.ap-southeast-2.amazo
naws.com/strapi-lms:latest
  ~/code/nyala/Building-APIs-with-Strapi  ⑂ main ±  docker push 3        .dkr.ecr.ap-southeast-2.amazonaws.com/strapi-lms:l
atest
The push refers to repository [3        .dkr.ecr.ap-southeast-2.amazonaws.com/strapi-lms]
ca8127bd835c: Pushed
672bf9e15278: Pushed
e47b622ee557: Pushed
38c5f2ffe0a5: Pushed
d181570fea5b: Pushed
bb9e16fb5863: Pushed
8dd06926952d: Pushed
b4b0971e42a6: Pushed
9df9c294d343: Pushed
722fd1e35a45: Pushed
100db43c90f1: Pushed
ffeece9978e9: Pushed
d44b6d1925f6: Pushed
f6d99686b622: Pushed
429965006bfe: Pushed
9b88fe065b35: Pushed
4ca605ea46de: Pushed
601f04850201: Pushed
846bd2f3b216: Pushed
2b3e667f5e92: Pushed
e891be0c59b2: Pushed
latest: digest: sha256:837b895ee47b0f37594302720bb8e26cb47fe8f9415a323c5ca68ce083f4d2ed size: 4719
```

Figure 10.12: Tagging and pushing the Docker image

Depending on your network connection speed, this process will take a couple of minutes to push the image to AWS.

Now that we have the Docker image on AWS, let's create a Fargate cluster to use our image.

Step 3 – Creating an AWS Fargate cluster

Before we start creating the cluster and containers, let's briefly describe the overall process and address some of the terminology that will be used in the process. The following diagram lists all the components that we will interact with:

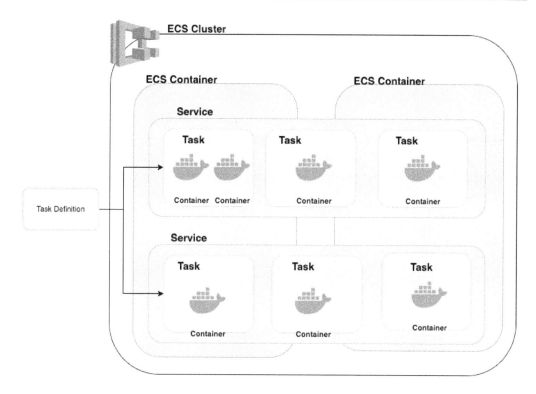

Figure 10.13: ECS objects diagram

The first step is to create a **Cluster**. A **Cluster** is a regional group of containers that we can use to run tasks. Next, we will create a **Task Definition**. You can think of the **Task Definition** as a blueprint of our application container. We will define which Docker image is to be used, the required amount of memory and CPU needed for the container, as well as environment variables that we want to pass to our container. Finally, we create a **Service**. The **Service** will use the **Task Definition** to run tasks. It will ensure that the minimum number of required/configured tasks are running all the time. In short, a **Service** is responsible for creating and running tasks.

> Note
>
> It is possible to skip the **Service** creation and run the tasks directly. However, this is ideal for a short-running job only, such as a CRON or batch job. For a long-running job, we should always use a service to manage and run the tasks.

Now that we have an overview of the process and components, let's get started:

1. Log in to the AWS console if you are not logged in, and navigate to the **Amazon Container Services** page.

2. Click on **Clusters** from the left-hand side menu, then click the **Create cluster** button.

3. Since we are going to run a Fargate cluster, choose **Networking Only** from the cluster template and then click the **Next step** button.

4. Choose a name for the cluster. We are going to use `strapi-lms`. Leave the remaining options as they are.

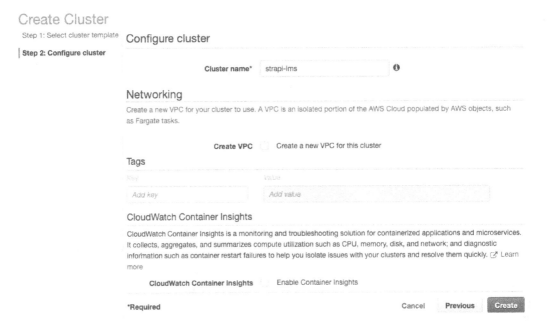

Figure 10.14: Creating an ECS cluster

5. Click the **Create** button to create the cluster. Once the cluster is created, click the **View Cluster** button to view the cluster information and return to the ECS page.

6. Next is the task definition. Click on the **Task Definitions** menu item from the left-hand side menu, then click the **Create new Task Definition** button.

7. Choose **FARGATE** from the launch type options, then click the **Next step** button.

8. Enter a name for the task definition. We will use `strapi-lms-task-def`. Leave the remaining settings as is and scroll down to the **Task execution IAM role** section.

9. This execution role allows you to pull images from ECR as well as publishing logs to the CloudWatch service. Choose **Create new role** to have a new role automatically created for us. The role name will be `ecsTaskExecutionRole`.

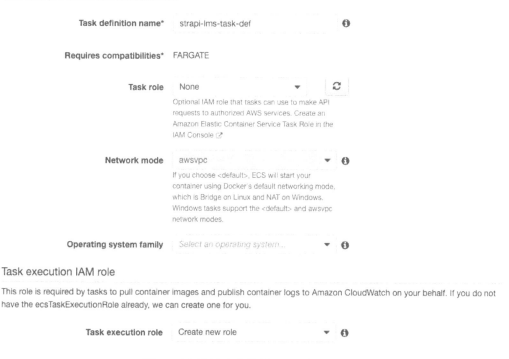

Figure 10.15: Task definition configuration

10. Next, we will need to configure the required memory and CPU for the task. We are going to use the minimum settings here with **0.5GB memory** and **0.25 vCPU**. These settings work fine for a demo or a small app; however, you should choose other options depending on your application requirements.

11. We will need to create the container definition now. Click on the **Add container** button under the **Container definitions** section.

12. Enter a name for the container. We will choose `strapi-lms-api`. For the **Image** field, enter the Docker image URI and make sure to specify the tag. If you do not have the Docker image URI, you can get it from the ECR page (refer to *Step 2 – Creating a Docker repository on AWS ECR*).

13. Scroll down to the **Port mappings** section and enter port `1337`, which is Strapi's default port.

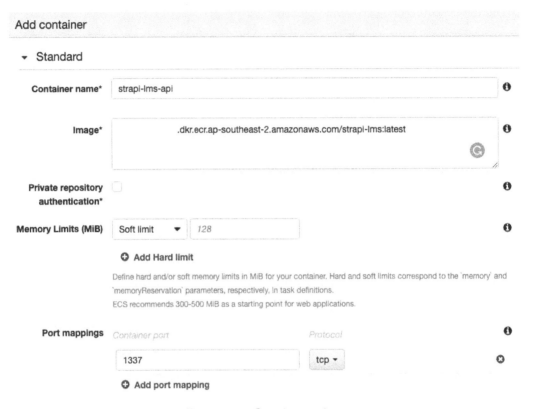

Figure 10.16: Container settings

14. Leave the remaining settings as is and scroll down to the **Environment** section. This is where we can define the `environment` variables needed by our app.

15. Add the required `environment` variables. For `DATABASE_URL`, we are going to reuse the same database we used with Heroku. However, feel free to create a new database on AWS if you prefer to.

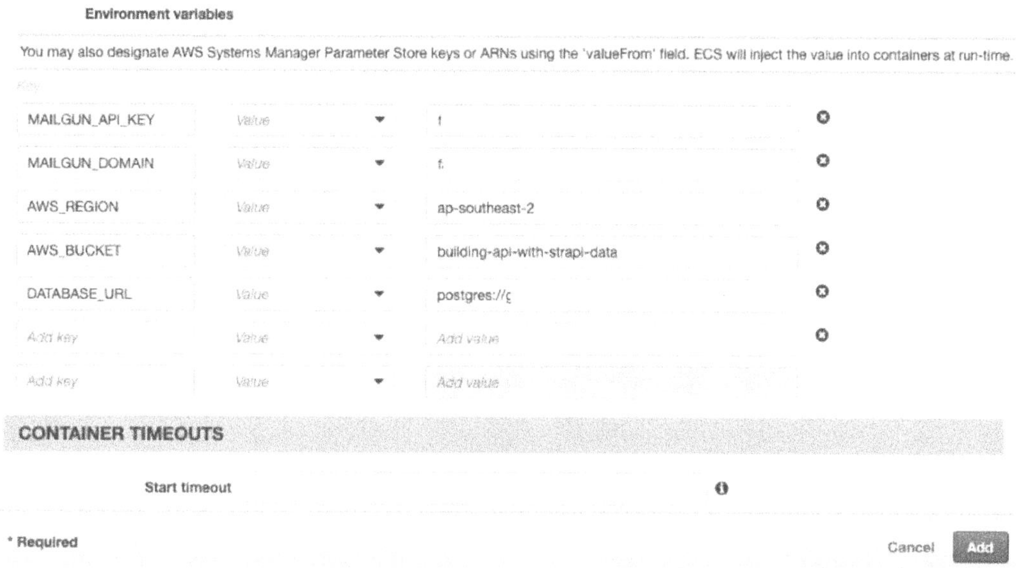

Environment variables

You may also designate AWS Systems Manager Parameter Store keys or ARNs using the 'valueFrom' field. ECS will inject the value into containers at run-time.

Key	Value		
MAILGUN_API_KEY	Value	▼	1
MAILGUN_DOMAIN	Value	▼	f.
AWS_REGION	Value	▼	ap-southeast-2
AWS_BUCKET	Value	▼	building-api-with-strapi-data
DATABASE_URL	Value	▼	postgres://c
Add key	Value	▼	Add value
Add key	Value	▼	Add value

CONTAINER TIMEOUTS

Start timeout   ❶

* Required   Cancel   **Add**

Figure 10.17: Container env variables

> **Note**
>
> When entering env variables, you have two options. Either use Value and
> enter the variable value directly or use ValueFrom to read the value from AWS
> Systems Manager Parameter Store. If you decide to use Parameter Store, make
> sure that the task execution role has permission to read from Parameter Store.

16. Click the **Add** button to add the container definition.

17. Leave the remaining settings as is, scroll down to the page bottom, and click the
    Create button to create the task definition.

18. The final step is to create a service to run our task. Return to the ECS main page and
    click on **Clusters** from the left-side menu.

19. Click on the cluster name we created earlier, in the **Services** tab, and click the
    Create button to create and deploy a new service.

20. Choose **FARGATE** as the launch type and make sure the task definition we created
    earlier is the selected one in the **Task Definition** list.

21. **Revision** should have only one value for now, **1 (latest)**. If we update the task
    definition, we will have a new revision number. This comes in handy if we decide to
    roll back our application to an older version, but for now, we will leave the value as is.

22. Enter a name for the service. We will use `strapi-lms-service`. Set **Number of tasks** to **1**. This will ensure that we always have at least one task running all the time.

Configure service

A service lets you specify how many copies of your task definition to run and maintain in a cluster. You can optionally use an Elastic Load Balancing load balancer to distribute incoming traffic to containers in your service. Amazon ECS maintains that number of tasks and coordinates task scheduling with the load balancer. You can also optionally use Service Auto Scaling to adjust the number of tasks in your service.

Figure 10.18: ECS service deployment configurations

23. Leave the remaining settings as is and click the **Next step** button to configure the network.

24. Choose your default VPC for the cluster VPC and select all available subnets.

25. By default, AWS blocks any inbound traffic. We will need to allow port `1337` to be able to access the dashboard and our API. Click the **Edit** button next to **Security group name**.

26. Make sure the **Create new security group** option is selected and update the name and description if you want.

27. You can remove the default rule that allows port `80` as we are not using this port.

28. Click the **Add rule** button, choose **Custom TCP** for the type, 1337 for the port range, and choose **Anywhere** for the source.

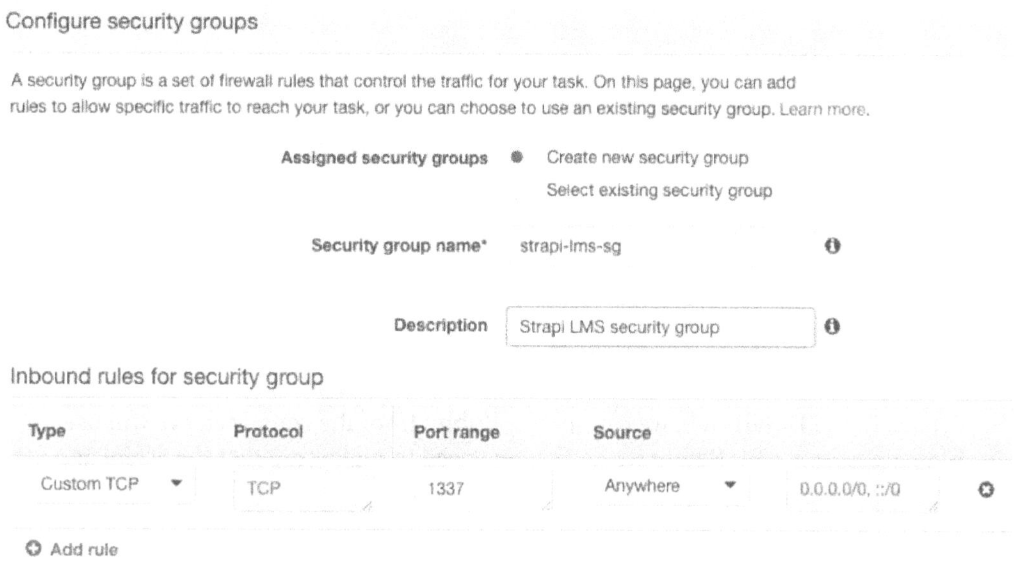

Configure security groups

A security group is a set of firewall rules that control the traffic for your task. On this page, you can add
rules to allow specific traffic to reach your task, or you can choose to use an existing security group. Learn more.

Assigned security groups	● Create new security group
	Select existing security group
Security group name*	strapi-lms-sg
Description	Strapi LMS security group

Inbound rules for security group

Type	Protocol	Port range	Source	
Custom TCP ▼	TCP	1337	Anywhere ▼	0.0.0.0/0, ::/0

⊕ Add rule

Figure 10.19: Networking settings

29. Click the **Save** button to save the security group settings.

30. Make sure the **Auto-assign Public IP** option is set to **Enabled**.

31. We will leave the remaining as is. Click the **Next** button.

32. We are not going to use a load balancer so we can skip this step by clicking the **Next** button.

33. The last page lists the configurations we chose. You can review them if needed, then click the **Create Service** button.

34. Once all the components are created, click the **View Service** button to view the service details.

35. Provisioning and running the task will take a few minutes. On the service detail page, click on the **Tasks** tab to view the status, we will need to wait till the status has changed to **RUNNING**.

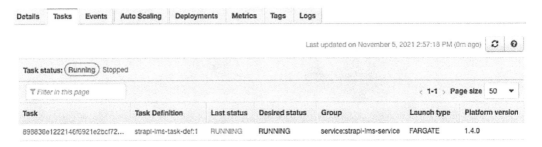

Figure 10.20: Task status

36. Once the task has the **RUNNING** status, click on the **Task** column to view the task details. In the **Details** tab, we can see the **Public IP** of the container. We will use this IP to access our API. Before we do that, we need to make sure that the Strapi app is running. Use the **Logs** tab to see Strapi startup logs.

37. Once we confirm that Strapi is up and running, use **Public IP** to access the dashboard. The URL should be `http://public_ip_here:1337/admin`.

At this stage, the API was successfully deployed to Fargate. Since we are using the same database that we used with Heroku, we should be able to log in to the dashboard using the same Super Admin credentials we used before.

Troubleshooting

Often, an invalid `environment` variable name or a missing AWS permission might cause our task to fail to start. If you have an issue running the container, you can investigate the logs to see where things went wrong. You can use the **Logs** tab to view the logs or use AWS CloudWatch. To use CloudWatch, do the following:

1. Navigate to the **CloudWatch** service page.

2. From the left-side menu, select **Log groups**.

3. There will be several log groups depending on how many services you are using or running. You can enter the task definition name in the search bar to filter the log groups.

4. Click on the task definition log group and select the most recent log stream or use the last event time to find the relevant log stream:

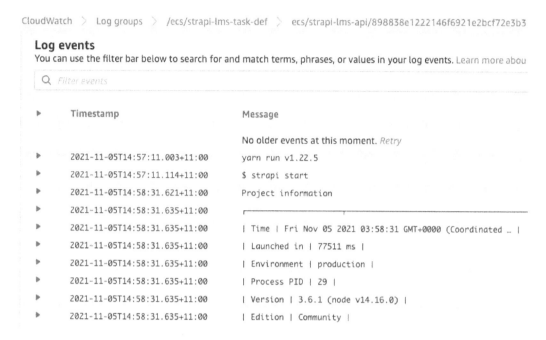

Figure 10.21: Task definition logs on CloudWatch

Once you have located the log stream, you can investigate the logs that went wrong.

Cleaning up

If you followed along and you do not need the task to be running anymore, it is better to stop it to avoid any unnecessary costs. We configured the minimum number of tasks when we created the service to be 1. This means if we stop the task manually, another one will start in a few minutes. Instead, let's change the service configuration, adjusting the minimum number of desired tasks to **0**:

1. Open the **strapi-lms** Cluster details page.
2. Click on **strapi-lms-service** to view the service details.
3. Click on the **Update** button to update the service.
4. Change the number of tasks to 0.
5. Save the changes by clicking the **Update service** button.

It will take a couple of minutes to stop the task. If you refresh the service details page, you should see the number of running tasks went down to zero.

Summary

In this chapter, we discussed Strapi application deployment, learned about deploying an API to a PaaS platform, as well as deploying it as a Docker container.

We started with a quick introduction to the Heroku platform. We then learned how to use the Heroku CLI to create a new application as well as to add a database to the application. Then we saw how simple it is to deploy to Heroku just by using the `git` command.

After that, we moved on to AWS and Docker container deployment. First, we talked about creating a Docker image for our API. Then, we learned how to push our Docker image to a private register we created on AWS ECR. Finally, we learned about ECS and saw how to create a cluster and deploy our API to that cluster.

In the next and final chapter, we will talk about testing the Strapi API using Jest. We will see how to prepare and configure the test environment, and how to write and run basic tests for the API.

11

Testing the Strapi API

Software testing is an important aspect of the software development life cycle—it helps in ensuring that the software is free from defects and that it meets the expected requirements. In this final chapter of the book, we will discuss testing the Strapi application. We will see which tools and libraries we need to test our Strapi **API** (short for **application programming interface**) and how to set up the testing environment. Finally, we will look at how to write and run tests against our code.

Here are the topics we will cover in this chapter:

- An overview of software testing
- Configuring test tools
- Setting up the test environment
- Writing and running tests

An overview of software testing

Before shipping the application to the end users, we will want to ensure that there are no bugs in it and that it will behave as expected. For example, we would not want to find out that the API is returning a different response from the expected one or that the server crashes when a specific input is used.

Ensuring that the application is bug-free is done through **software testing**. Software testing is a process of evaluating that the software will function and operate as expected without any surprises. The following diagram shows a testing pyramid. A testing pyramid is an approach to structure the test suites:

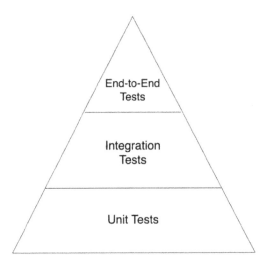

Figure 11.1: Example of software-testing pyramid

Unit testing focuses on testing small units of the application, such as a class, function, or algorithm. **Integration testing** focuses on a small number of modules and tests how they work together. **End-to-end (E2E)** testing, as the name suggests, focuses on testing the entire application flow from beginning to end.

As the main focus of this book is not software testing, we will only focus on unit testing. However, before we start testing our code, we will need to set up and configure tools to help us achieve our task, so let's do so in the next section.

Configuring test tools

The first thing we need to do is install the required tools necessary for testing our Strapi application. The first library we need is **Jest**. Jest is a JavaScript library developed and maintained by Facebook; it is mainly used for unit testing. Let's get started, as follows:

1. To install Jest, launch your terminal and execute the following command:

   ```
   yarn add -D jest
   ```

   This command will install Jest for us. Notice that we have used the -D flag when installing the library. This flag is used to indicate that the library we are about to install is a development library. We only need it while we are developing and working on the application, but we do not need it in the production environment since tests are run and executed before we package and ship the application.

   Jest is great to test the functionality of our API components such as controller logic or utility functions, but we cannot use it to test **HyperText Transfer Protocol** (**HTTP**) requests.

2. To test our API endpoints using HTTP requests, we will need to install a library called **Supertest**. In the terminal, execute the following command:

   ```
   yarn add -D supertest
   ```

   Supertest is an HTTP assertion library that allows us to test the Node.js HTTP server; we will be using it together with Jest.

   > **Note**
   >
   > If you have not been following along from the beginning, you will need to install `sqlite3` if you do not have it. From your terminal, run the `yarn add -D sqlite` command to install `sqlite3`.

3. After we have installed the required libraries, let's create a script that we will use to run our tests later on, then open `package.json`, and add the following test script to the `scripts` section. The test script simply calls Jest to run and execute all test files in our project:

   ```
   "test": "jest --detectOpenHandles"
   ```

   The updated `scripts` section should now look like this:

   ```
   "scripts": {
       "develop": "strapi develop",
       "start": "strapi start",
   ```

```
    "build": "strapi build",
    "strapi": "strapi",
    "db:start": "docker compose -f docker-compose-
    dev.yml up -d",
    "db:stop": "docker compose -f docker-compose-
    dev.yml down",
    "db:logs": "docker logs -f strapi_api",
    "test": "jest --forceExit --detectOpenHandles"
}
```

The test script we just added will execute jest in our project, which will try to find all test files and execute the tests in them. If we run the yarn test command now, we will have dozens of errors. The reason for those errors is that Jest will look for any test files in the project, including the node_modules and the .cache folders. We should exclude those folders from our tests because the packages in the node_modules folder should have been tested by their author before releasing them, and the .cache folder contains files from Strapi itself that we do not need to test.

4. To exclude those folders from our test, add the following code to the end of the package.json file:

```
"jest": {
    "testPathIgnorePatterns": [
    "/node_modules/",
    ".cache"
    ],
    "testEnvironment": "node",
    "testTimeout": 15000
}
```

This will let Jest know that we want to exclude the node_modules and .cache folders from our tests; so, when Jest is run, it will not look inside those folders and will simply ignore them. testEnvironment is there to let Jest know that we are running a Node.js environment. The last line adjusts the default Jest timeout from 5 seconds to 15 seconds (the value is expressed in **milliseconds (ms)**); this is important to avoid timeout errors later on when we run our tests.

To test our setup, let's create a simple test file, as follows:

1. Create a new folder called tests at the root of the project. We will put all of our tests in this folder later on.

2. Create a simple `app.test.js` test file. Notice that we add `.test` to the filename before the extension. This is to let Jest know that this is a test file and should be treated as such.

3. Add a simple test to the file, as follows:

```
test("Setup is working", () => {
    console.log("Hello from test file");
});
```

This simple test file has a `"Setup is working"` single test, and it does nothing at the moment other than just print a message to the console.

4. Save the changes to the file, and then run the `yarn test` command from your terminal, as illustrated in the following screenshot:

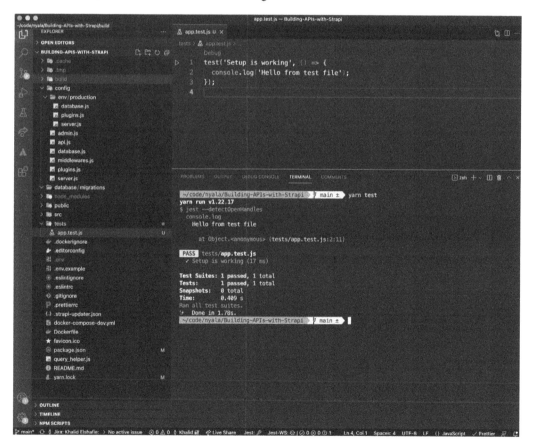

Figure 11.2: Example of running tests

You should see a `Hello from test file` message in the console and that our dummy test has passed.

Now that we have installed the required libraries to test our Strapi API, let's set up the testing environment next.

Setting up the test environment

We have installed the required libraries to write and run the tests, so the next thing on our list is to prepare the testing environment itself. There are two main things we need to do here—the first is to prepare the database configurations for the test database, and the second thing is to prepare the Strapi instance that is required for the tests.

Preparing the database configuration

When we are running our tests, we will often want to interact with the database. We will not want to use the real database for sure to run our tests, and the development database might contain old or dirty data that could potentially break the tests. Instead, we will set up a temporary database for the tests that we can use while running the tests so that they are run on a clean database, then remove it later on when it isn't needed.

When we use `jest` to run the tests, it will set the `NODE_ENV` variable value to `test`. This means we can easily create a configuration for the test environment by placing the configuration file in the `config/env/test` folder, so let's do that now, as follows:

1. In the `config/env` directory, create a new folder called `test`.

2. Create a new `config/env/test/database.js` file. We will use this file to set up the test database information.

3. Add the following code to the file:

```
const path = require('path');

module.exports = ({ env }) => ({
  connection: {
    client: 'sqlite',
    connection: {
      filename: env('DATABASE_FILENAME','.tmp/test.db'),
    },
    useNullAsDefault: true,
    pool: {
```

```
      min: 0,
      max: 1,
    },
  },
});
```

This setting will use a `test.db` database whenever we are running the tests or whenever the `NODE_ENV` environment variable value is set to `test`. Next, let's configure the Strapi instance.

Preparing the Strapi instance

In order for us to be able to test anything that is Strapi-related, we will need to have a Strapi instance defined in the test environment. Let's do that now, as follows:

1. Create a new `helpers` folder in the `tests` directory.
2. Create a new file called `strapi.js`. The full path to this file should be `tests/helpers/strapi.js`.
3. Add the following code to it:

    ```
    const Strapi = require('@strapi/strapi');

    let instance;
    const setupStrapi = async () => {
        if (!instance) {
          instance = await Strapi().load();
          instance.server.mount();
        }
        return instance;
    };

    module.exports = { setupStrapi };
    ```

The previous code declares a single function called `setupStrapi`. This function uses a **singleton** design pattern to initialize the Strapi instance. This instance will be available in our tests by simply calling `strapi` when needed.

> **What is a Singleton?**
>
> A singleton is a software design pattern used for object creation. It ensures that a class has only one instance.

Our setup is complete now, so let's create a simple test to ensure that everything is properly configured, as follows:

1. Let's go back to our `tests/app.test.js` file.

2. Replace the test content with the following code:

    ```
    const { setupStrapi } = require("./helpers/strapi");

    /** this code is called once before any test is called */
    beforeAll(async () => {
      await setupStrapi();
    });

    test("strapi is defined", () => {
    expect(strapi).toBeDefined();
    });
    ```

    This simple test file exposes a single test titled `strapi is defined`. It makes use of the `setupStrapi` helper we created earlier and tests that the `strapi` instance is defined. In the `beforeAll` function, we call the `setupStrapi` function to initialize the Strapi instance if it was not initialized.

    > **What does the beforeAll function do?**
    >
    > `beforeAll` is a helper function provided by Jest to help in setting up tests. As the name suggests, the `beforeAll` function is executed once by Jest before running all tests. There is also an `afterAll` function that is executed once after all the tests are run. Additionally, there are also `beforeEach` and `afterEach` helpers, and those are executed before and after each test.

3. Save the changes to the file, then run the `yarn test` command from the terminal to execute the tests, as illustrated in the following screenshot:

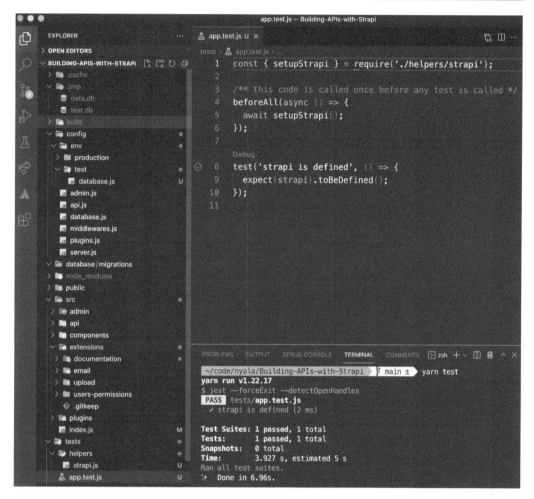

Figure 11.3: Testing if the Strapi instance is defined

Windows Users

The following steps will not work for Windows users. The reason for that is Windows will lock the SQLite database file (test.db), and if we attempt to delete it, we will get an error. If you are running on Windows, you will need to remove the file manually.

4. The last thing we need to set up is a teardown script that will remove the test database for us after all tests are run. This will allow us to have a fresh and clean database every time we run our tests. In the package.json file, add the globalTeardown option to the jest block, as follows:

```
"jest": {
    "testPathIgnorePatterns": [
    "/node_modules/",
    ".cache"
    ],
    "testEnvironment": "node",
    "testTimeout": 15000,
    "globalTeardown": <rootDir>/tests/cleanup.js
}
```

The globalTeardown option allows us to specify a module that exports a function that is triggered once after all test suites.

5. Next, create a cleanup.js file in the tests folder and add the following code to it:

```
const fs = require('fs');
const os = require('os');

const { setupStrapi } = require('./helpers/strapi');
const cleanup = async () => {
    await setupStrapi();
    const dbSettings = strapi.config.get('
    database.connection.connection');

    //close server to release the db-file
    await strapi.destroy();

    // Check if we are on windows
    const isWindows = os.platform() === 'win32';
    //delete test database after all tests
    if (!isWindows && dbSettings &&
    dbSettings.filename) {
        if (fs.existsSync(dbSettings.filename)) {
```

```
        fs.unlinkSync(dbSettings.filename);
      }
    }
  };


  module.exports = cleanup;
```

6. Run the `yarn test` command one more time. After the tests are run, the `.tmp/` `test.db` file should be deleted automatically.

At this stage, we have everything configured and set up our Strapi API for testing, so let's do that in the next section.

Writing and running tests

We are going to give a couple of examples of writing tests. First, we will use a public endpoint and see how we can write a test against our `public` routes, then we will see how to test a secure endpoint.

Testing a public endpoint

We will start by testing a public endpoint first; let's consider the `GET /api/` `classrooms` endpoint. We want to test the endpoint is working as expected—that is, it is accessible and returning the expected result.

Before we start writing the tests, let's make sure that we are allowing public access to this endpoint by updating the permissions. Update the `bootstrap` function in the `src/` `index.js` file, allowing `public` access to the classroom's `find` endpoint, as illustrated in the following code snippet:

```
await enablePermission("public", "classroom","classroom",
"find");
```

We should be able to access the `GET /api/classrooms` endpoint without the need to authenticate. Now that the permission is set up, let's write our test, as follows:

1. Create a new `classroom.test.js` file in the `tests` folder.
2. Add the following code to the file:

```
const request = require("supertest");
const { setupStrapi } = require("./helpers/strapi");
```

```
describe("classroom controller test", () => {
  beforeAll(async () => {
   await setupStrapi();
  });

  afterAll(() => {
     Strapi.server.destroy();
  });

  it("should return list of classroom", async () => {
   await request(strapi.server.httpServer)
     .get("/api/classrooms")
     .expect(200)
     .expect("Content-Type", /json/);
  });
});
```

We start by requiring the superset and setupStrapi helpers. Then, we use beforeAll and afterAll to set up a Strapi instance before all tests are run and destroy it once all tests are finished. We then defined a test titled should return list of classroom. In our test, we used Supertest to issue a GET request to the /api/classrooms endpoint and then used expect to assert that the endpoint will return a 200 OK status code and that the content-type of the response is **JavaScript Object Notation (JSON)**, using the /json/ **regular expression (regex)**.

What Does describe("classroom controller test") Do?

The describe function allow us to group related test together. When we use describe, we create a block that we can use to group all related tests together. A single test file, referred to as a *test suite*, can have multiple describe blocks, and each block can contain multiple tests.

---

3. Run the tests using the yarn test command:

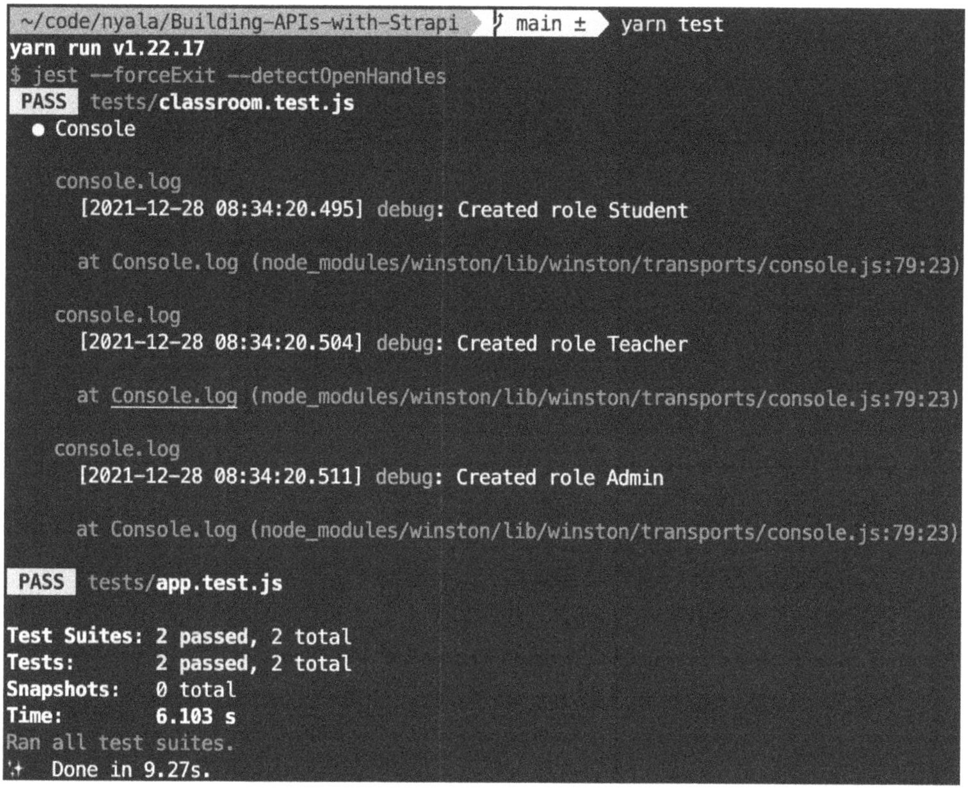

```
~/code/nyala/Building-APIs-with-Strapi    main ±    yarn test
yarn run v1.22.17
$ jest --forceExit --detectOpenHandles
 PASS  tests/classroom.test.js
  ● Console

    console.log
      [2021-12-28 08:34:20.495] debug: Created role Student

      at Console.log (node_modules/winston/lib/winston/transports/console.js:79:23)

    console.log
      [2021-12-28 08:34:20.504] debug: Created role Teacher

      at Console.log (node_modules/winston/lib/winston/transports/console.js:79:23)

    console.log
      [2021-12-28 08:34:20.511] debug: Created role Admin

      at Console.log (node_modules/winston/lib/winston/transports/console.js:79:23)

 PASS  tests/app.test.js

Test Suites: 2 passed, 2 total
Tests:       2 passed, 2 total
Snapshots:   0 total
Time:        6.103 s
Ran all test suites.
✨  Done in 9.27s.
```

Figure 11.4: Running classroom controller test

If everything is set up correctly, we should see that two tests passed—one is for testing if Strapi is defined, and the other test is the one we just created to test the classroom endpoint.

Even though our classroom test endpoint passed, it is still missing something. Our test is simply calling the endpoint and making sure that we get a 200 response, while our test name indicates we are testing that we are getting a list of classrooms.

Let's update our test to check that we are getting the classrooms in the response. Update the test file as follows:

```
const request = require("supertest");
const { setupStrapi } = require("./helpers/strapi");

describe("classroom controller test", () => {
  beforeAll(async () => {
    await setupStrapi();
```

```
  });

  afterAll(() => {
    Strapi.server.destroy();
  });

  it("should return list of classroom", async () => {
    await request(strapi.server.httpServer)
      .get("/api/classrooms")
      .expect(200)
      .expect("Content-Type", /json/)
      .then(response => {
        const { body } = response;
        expect(body).toBeDefined();
        expect(body).toHaveProperty("data");
        const { data } = body;
        expect(data.length).toBe(5);
        expect(data[0].attributes.name).toBe("classroom_1");
      });
  });
});
```

The updated test now checks that the response body is defined, we have 5 classrooms returned in the response, and that the first `classroom` name is `classroom_1`. If we run this test now, it will fail because our test database is empty and we do not have any records in the database yet.

To populate the test database with sample data for our tests, let's create a simple utility that will seed the test database, as follows:

1. Create a new file called `data.helper.js` in the `tests/helpers` folder.

2. Create a `createTestClasses` function and add the following code to it:

```
const createTestClasses = async (numberOfClasses = 10) =>
{
  try {
    const classroomsPromise = [];
    const min = 1;
```

```
  const max = 30;
  Array(numberOfClasses)
   .fill(null)
   .forEach((_item, index) => {
     const name = `classroom_${index + 1}`;
     const maxStudents = Math.random() * (max - min + 1)
     + min;
     classroomsPromise.push(
       strapi.services("
       api::classroom.classroom").create({
         data: {
           name,
           description: `Description of the classroom
           ${name}`,
           maxStudents: Math.floor(maxStudents)
         }
       })
     );
   });

  await Promise.all(classroomsPromise);
} catch (e) {
  throw new Error("Failed to create mock data");
}
};

module.exports = {
  createTestClasses,
};
```

You might recognize this code from *Chapter 6, Dealing with Content*. We used this code in our initial seed to create sample data in the database. We can use it here again to populate the test database with sample data to work with while running the tests.

3.　Now that we have a `helper` class to generate sample data, go ahead and update our `classroom` controller test by adding a `beforeAll` Hook to create data before all tests are executed, as follows:

```
const request = require("supertest");
const { setupStrapi } = require("./helpers/strapi");

const {createTestClasses} = require("../helpers/data.
helper");
const numberOfClasses = 5;
describe("classroom controller test", () => {
  // Run before all tests, create dummy classes
  beforeAll(() => {
    await setupStrapi();
    await createTestClasses(numberOfClasses);
  });

  it("should return list of classroom", async () => {
    await request(strapi.server.httpServer)
      .get("/api/classrooms")
      .expect(200)
      .expect("Content-Type", /json/)
      .then(response => {
        const { body } = response;
        expect(body).toBeDefined();
        expect(body).toHaveProperty("data");
        const { data } = body;
        expect(data.length).toBe(numberOfClasses);
        expect(data[0].attributes.name).
        toBe("classroom_1");
      });
  });
});
```

4. Run the tests again using the `yarn test` command. We should now see that our tests are passing.

With this example, we were able to test that our `classroom` endpoint is accessible and returning the correct response code. We were also able to test the response itself and validate that we are getting data. However, the endpoint we were testing was public, but not all our API's endpoints are public, and the majority of them require the user to be logged in. Let's discuss in the next section how we can run our tests against a secure endpoint.

Testing a secure endpoint

In this example, we will see how we can test an endpoint that requires the user to be logged in. We will use the `create classroom` endpoint to illustrate this scenario—this will also help us see how we can send POST requests with a payload in our tests. Let's get started, as follows:

1. Open the `tests/classroom.test.js` classroom test file.

2. Add a new test. Let's call it `should create new classroom`, with the following body:

```
it("should create new classroom", async () => {
  // classroom we will create
  const classroom = {
    name: "Test classroom",
    description: "Class room for unit tests",
    maxStudents: 10,
  };

  const payload = {
    data: {
      ...classroom
    }
  };

  await request(strapi.server.httpServer)
    .post("/api/classrooms")
    .set("Content-Type", "application/json")
    .send(payload)
    .expect(200)
```

```
    .then((response) => {
        const { body } = response;
        expect(body).toBeDefined();
        expect(body).toHaveProperty("data")

        const { data } = body;
        expect(data.id).toBeDefined();
        expect(data.attributes.name).toBe(classroom.name);
        expect(data.attributesdescription).toBe(classroom.
        description);
        expect(data.attributes.maxStudents).toBe(classroom.
        maxStudents);
    });
});
```

We started the test by defining an object called classroom that holds the information of the classroom we want to create. Then, we send the POST request to the /classrooms endpoint. Note that we set the Content-Type header to application/json as our API expects that from us, and we used the send method to set our request body.

Finally, we defined our expectations—that the API will return 200 and the body will contain an id value for the new classroom, and the name, description, and maxStudent classroom properties will match what we sent in the request body.

3. If we run the test now, it will fail with a 403 Forbidden response code. This is expected since that endpoint expects a user to be logged in. You can see the output in the following screenshot:

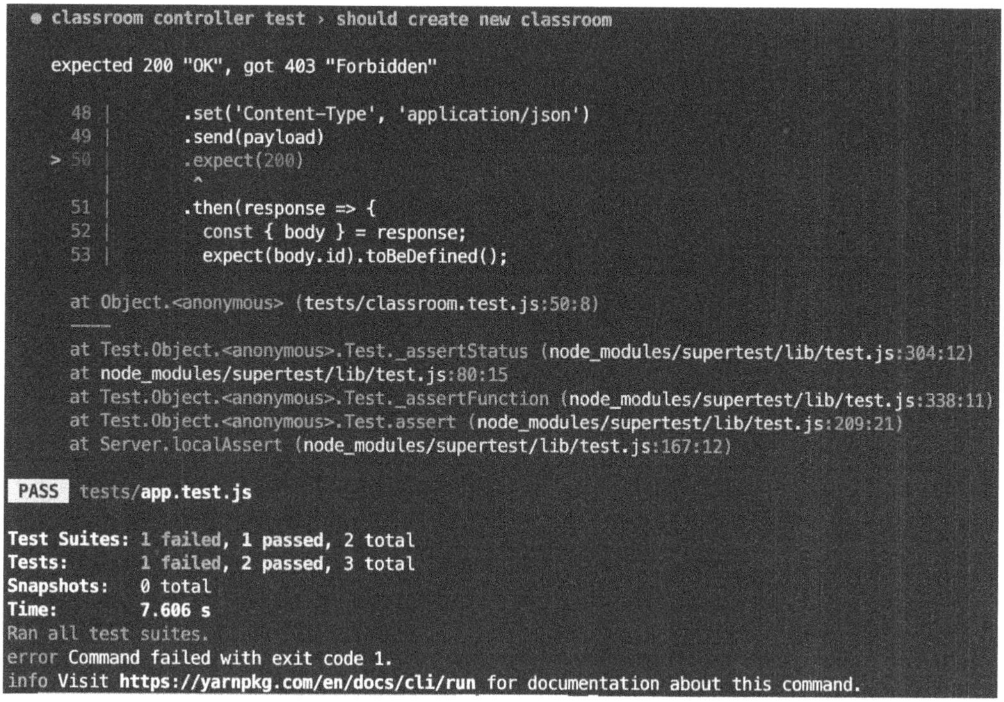

```
● classroom controller test › should create new classroom

  expected 200 "OK", got 403 "Forbidden"

     48 |         .set('Content-Type', 'application/json')
     49 |         .send(payload)
   > 50 |         .expect(200)
        |          ^
     51 |         .then(response => {
     52 |           const { body } = response;
     53 |           expect(body.id).toBeDefined();

  at Object.<anonymous> (tests/classroom.test.js:50:8)
  ─────
  at Test.Object.<anonymous>.Test._assertStatus (node_modules/supertest/lib/test.js:304:12)
  at node_modules/supertest/lib/test.js:80:15
  at Test.Object.<anonymous>.Test._assertFunction (node_modules/supertest/lib/test.js:338:11)
  at Test.Object.<anonymous>.Test.assert (node_modules/supertest/lib/test.js:209:21)
  at Server.localAssert (node_modules/supertest/lib/test.js:167:12)

 PASS  tests/app.test.js

Test Suites: 1 failed, 1 passed, 2 total
Tests:       1 failed, 2 passed, 3 total
Snapshots:   0 total
Time:        7.606 s
Ran all test suites.
error Command failed with exit code 1.
info Visit https://yarnpkg.com/en/docs/cli/run for documentation about this command.
```

Figure 11.5: Failed to create classroom when user is not authenticated

To solve this problem, we will need to send a **JSON Web Token** (**JWT**) in the request header. Remember that our test database is created every time we run tests and dropped at the end. We will start by creating a user in the database; then, we can use that user to issue a JWT that we can use to access a secure endpoint.

4. Let's add a function called `createUser` to our `tests/helpers/data. helper.js` file, as follows:

```
const createUser = async (type = 'admin', username =
'testuser') => {
    try {
    // Get the role from the database
    const role = await strapi.db.query('plugin::users-
    permissions.role').findOne({ where: { type } });

    // Create the user
    return await strapi.db.query('
     plugin::users-permissions.user').create({
      data: {
```

```
        username,
        email: `${username}@strapi.com`,
        password: 'password',
        provider: 'local',
        confirmed: true,
        role: role ? role.id : null,
      },
    });
  } catch (e) {
    throw new Error('Failed to create mock user');
  }
};
```

Our `createUser` function accepts two optional parameters—a role defaulted to `admin` and a username defaulted to `testuser`. This will give us a bit of flexibility when we create users in the future when our tests evolve.

We started by querying the database for the role. Remember that the database is seeded with three roles (covered in *Chapter 9, Production-Ready Applications*) every time we run our tests. Then, we create a user with that role and return the result.

5. Don't forget to export the function so that we can use it in our tests. Here's the code to achieve this:

```
module.exports = {
  createTestClasses,
  createUser,
};
```

6. Now that we have a `createUser` helper, let's switch back to our `tests/classroom/index.js` file. First, import the newly created function, as follows:

```
const { createTestClasses, createUser } = require("../
helpers/data.helper");
```

7. Then, let's update our `should create new classroom` test. Let's create a user and issue a JWT, then add the following code right before our request call:

```
//Create user
const testUser = await createUser();
// issue JWT for that user
```

```
const jwt = await strapi.plugins["users-permissions"].
services.jwt.issue({
  id: testUser.id,
});
```

8. Now that we have the JWT, we can pass it in our request. Our updated test should now look like this:

```
it("should create new classroom", async () => {
  // classroom we will create
  const classroom = {
   name: "Test classroom",
   description: "Class room for unit tests",
   maxStudents: 10,
  };

  const payload = {
   data: {
    ...classroom
   }
  };

  const testUser = await createUser();
  const jwt = await strapi.plugins["users-
  permissions"].services.jwt.issue({
   id: testUser.id
});

  await request(strapi.server.httpServer)
    .post("/api/classrooms")
    .set("Content-Type", "application/json")
    .set("Authorization", `Bearer ${jwt}`)
    .send(payload)
    .expect(200)
    .then((response) => {
      const { body } = response;
      expect(body).toBeDefined();
```

```
      expect(body).toHaveProperty("data")


    const { data } = body;
    expect(data.id).toBeDefined();
    expect(data.attributes.name).toBe(classroom.name);
    expect(data.attributesdescription).toBe(classroom.
    description);
    expect(data.attributes.maxStudents).toBe(classroom.
    maxStudents);
  });
});


it("should create new classroom", async () => {
  // classroom we will create
  const classroom = {
   name: "Test classroom",
   description: "Class room for unit tests",
   maxStudents: 10,
  };


  const testUser = await createUser();


  const jwt = await strapi.plugins["users-permissions"].
  services.jwt.issue({
   id: testUser.id,
});


await request(strapi.server)
  .post("/classrooms")
  .set("Content-Type", "application/json")
  .set("Authorization", `Bearer ${jwt}`)
  .send(classroom)
  .expect(200)
  .then((response) => {
    const { body } = response;
    expect(body.id).toBeDefined();
    expect(body.name).toBe(classroom.name);
```

```
        expect(body.description).toBe(
        classroom.description);
        expect(body.maxStudents).toBe(
        classroom.maxStudents);
      });
    });
```

9. Let's run the `yarn test` command again. This time, our test will pass.

We are now able to test both public and secure endpoints. We will end this chapter with the remark that software testing is a huge topic that probably needs a dedicated book on its own rather than just a chapter in a book. However, we hope that this chapter helps you in getting started with testing your application and code.

Summary

In this chapter, we explored how we can test our Strapi API, and started with a brief introduction to software testing. Then, we started setting up our environment and installed the required libraries for our test.

After we had our environment ready, we started by testing a public endpoint. We used the `GET /api/classrooms` endpoint as an example and saw how we can issue an HTTP request to that endpoint and how we can assert that we are getting the expected result.

Finally, we explored how we can test a secure endpoint that requires the user to be authenticated. We created a helper function that created a user with a specific role, then we used that user to issue a JWT token. Afterward, we used the token in our tests, passing it along in the request header, and we were able to test a secure endpoint.

We have covered a variety of topics in this book; technologies, techniques, and best practices, as well as briefly touching on cloud-related topics. We do not expect anybody to take in all of this in one read. Therefore, when setting forth to develop your own Strapi-based API and backend, we hope this book will serve as a reference and a great foundation!

Appendix: Connecting a React App to Strapi

We will describe how we can connect a simple frontend app to the API we have built throughout the book. We will use a sample React app as our frontend app.

About the React app

The sample app we will use here is a simple and minimal app built with ReactJS to illustrate how can we hook our API to a frontend app.

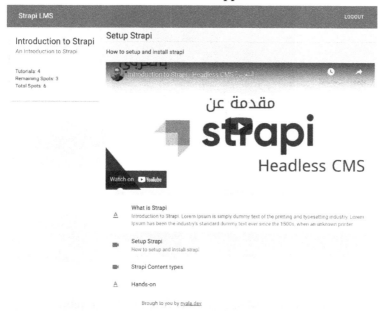

Figure 1: Example of the frontend app after startup

The React app allows students to log in, see a list of classrooms, enroll in a classroom, and see the classroom tutorials. The tech stack for the app is as follows:

- **ReactJS (TypeScript)**
- **Reach Router**: The routing component
- **Material UI**: Used to build the UI components
- **Axios**: HTTP library to communicate with our Strapi API
- **ReactPlayer**: A React library to work with YouTube videos

Understanding the React app folder structure

The React app is hosted on GitHub, and you can clone it from the following GitHub repository:

`https://github.com/NyalaDev/strap-lms-react`

Once you have cloned it, you can open it with your text editor. The following is the folder structure:

Figure 2: Frontend app folder structure

The app uses a common React app structure. All components reside within the `src` directory and are categorized into the following:

- `common`: The `common` directory contains constants files; it defines constants that are used across the app.

- **components**: The `components` directory holds various UI components used in the app.

- **context**: The application uses React `context` to manage the application auth state.

- **hooks**: Custom Hooks are used to handle authentication.

- **pages**: A top-level component, this contains the **login** page, **home** page, and **classroom detail** page.

- **services**: There are two services used by the app: the local storage service is used to work with the local storage and the API service is used to define all API calls to our Strapi API.

- **types**: Contains TypeScript type definitions used in the app.

- **index.tsx**: `index.tsx` is the main entry point and it contains definitions for all routes used in the app.

Running the app

To run the sample app, first we will need to create a configuration file to set the API URL, then we can run the app using the start script:

1. Create a `.env.local` file in the project root.

2. Add the following to the new file: `REACT_APP_STRAPI_API_URL=http://localhost:1337`.

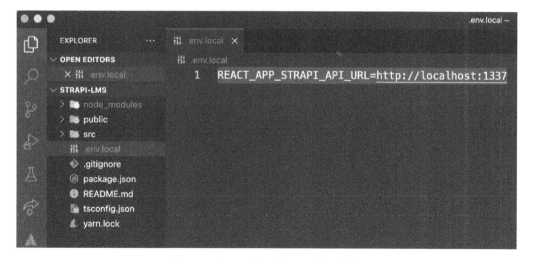

Figure 3: Example of the .env.local file

The `REACT_APP_STRAPI_API_URL` env variable points to the Strapi API we created in the book. This environment variable is referenced and used in `services/api.ts`. If you are running your Strapi instance on a different machine, make sure to adjust the value of this variable to point to your Strapi application.

3. Save the changes to the file, and then from the terminal run the command `yarn start` to launch the development server. Once the server is up and running, a new browser tab should atomically open. If not, open your browser and navigate to `http://localhost:3000`. You will be presented with the login screen:

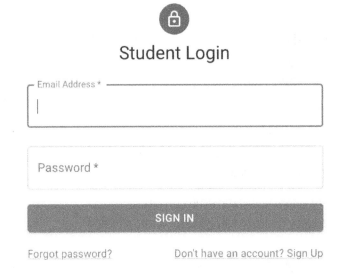

Figure 4: Frontend app login page

Make sure your Strapi instance is running. Use a student account to log in to the app and explore the dashboard. Feel free to make any changes you want.

Index

Y

Packt.com

Subscribe to our online digital library for full access to over 7,000 books and videos, as well as industry leading tools to help you plan your personal development and advance your career. For more information, please visit our website.

Why subscribe?

- Spend less time learning and more time coding with practical eBooks and Videos from over 4,000 industry professionals

- Improve your learning with Skill Plans built especially for you

- Get a free eBook or video every month

- Fully searchable for easy access to vital information

- Copy and paste, print, and bookmark content

Did you know that Packt offers eBook versions of every book published, with PDF and ePub files available? You can upgrade to the eBook version at packt.com and as a print book customer, you are entitled to a discount on the eBook copy. Get in touch with us at customercare@packtpub.com for more details.

At www.packt.com, you can also read a collection of free technical articles, sign up for a range of free newsletters, and receive exclusive discounts and offers on Packt books and eBooks.

Other Books You May Enjoy

If you enjoyed this book, you may be interested in these other books by Packt:

UI Testing with Puppeteer

Dario Kondratiuk

ISBN: 978-1-80020-678-6

- Understand browser automation fundamentals
- Explore end-to-end testing with Puppeteer and its best practices
- Apply CSS Selectors and XPath expressions to web automation
- Discover how you can leverage the power of web automation as a developer
- Emulate different use cases of Puppeteer such as network speed tests and geolocation

API Testing and Development with Postman

Dave Westerveld

ISBN: 978-1-80056-920-1

- Find out what is involved in effective API testing
- Use data-driven testing in Postman to create scalable API tests
- Understand what a well-designed API looks like
- Become well-versed with API terminology, including the different types of APIs
- Get to grips with performing functional and non-functional testing of an API
- Discover how to use industry standards such as OpenAPI and mocking in Postman

Packt is searching for authors like you

If you're interested in becoming an author for Packt, please visit authors. packtpub.com and apply today. We have worked with thousands of developers and tech professionals, just like you, to help them share their insight with the global tech community. You can make a general application, apply for a specific hot topic that we are recruiting an author for, or submit your own idea.

Hi!

We're Khalid and Mozafar, the authors of *Designing Web APIs with Strapi*. We really hope you enjoyed reading this book and found it useful for increasing your productivity and efficiency in building APIs with Node.js and Strapi.

It would really help us (and other potential readers!) if you could leave a review on Amazon sharing your thoughts on *Designing Web APIs with Strapi*.

Go to the link below or scan the QR code to leave your review:

https://packt.link/r/180056063X

Your review will help us to understand what's worked well in this book, and what could be improved upon for future editions, so it really is appreciated.

Best Wishes,

Khalid Elshafie

Mozafar Haider